Mathematical
Olympiad
in China
Problems and Solutions

Mathematical 数Olympiad in China

Problems and Solutions

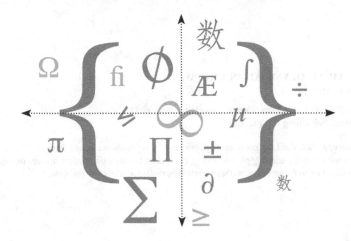

Editors

Xiong Bin
East China Normal University, China

Lee Peng Yee
Nanyang Technological University, Singapore

East China Normal
University Press

World Scientific

Published by

East China Normal University Press
3663 North Zhongshan Road
Shanghai 200062
China

and

World Scientific Publishing Co. Pte. Ltd.
5 Toh Tuck Link, Singapore 596224
USA office: 27 Warren Street, Suite 401-402, Hackensack, NJ 07601
UK office: 57 Shelton Street, Covent Garden, London WC2H 9HE

British Library Cataloguing-in-Publication Data
A catalogue record for this book is available from the British Library.

MATHEMATICAL OLYMPIAD IN CHINA
Problems and Solutions

ISBN-13 978-981-270-789-5 (pbk)
ISBN-10 981-270-789-1 (pbk)

Printed in Singapore.

Editors

XIONG Bin *East China Normal University, China*

LEE Peng Yee *Nanyang Technological University, Singapore*

Original Authors

MO Chinese National Coaches Team of 2003 – 2006

English Translators

XIONG Bin *East China Normal University, China*

FENG Zhigang *Shanghai High School, China*

MA Guoxuan *East China Normal University, China*

LIN Lei *East China Normal University, China*

WANG Shanping *East China Normal University, China*

ZHENG Zhongyi *High School Affiliated to Fudan University, China*

HAO Lili *Shanghai Qibao Senior High School, China*

WEE Khangping *Nanyang Technological University, Singapore*

Copy Editors

NI Ming *East China Normal University Press, China*

ZHANG Ji *World Scientific Publishing Co., Singapore*

XU Jin *East China Normal University Press, China*

Preface

The first time China sent a team to IMO was in 1985. At that time, two students were sent to take part in the 26th IMO. Since 1986, China has always sent a team of 6 students to IMO except in 1998 when it was held in Taiwan. So far (up to 2006), China has achieved the number one ranking in team effort for 13 times. A great majority of students have received gold medals. The fact that China achieved such encouraging result is due to, on one hand, Chinese students' hard working and perseverance, and on the other hand, the effort of teachers in schools and the training offered by national coaches. As we believe, it is also a result of the educational system in China, in particular, the emphasis on training of basic skills in science education.

The materials of this book come from a series of four books (in Chinese) on *Forward to* IMO: *a collection of mathematical Olympiad problems* (2003 - 2006). It is a collection of problems and solutions of the major mathematical competitions in China, which provides a glimpse on how the China national team is selected and formed. First, it is the China Mathematical Competition, a national event, which is held on the second Sunday of October every year. Through the competition, about 120 students are selected to join the China Mathematical Olympiad (commonly known as the Winter Camp), or in short CMO, in January of the second year. CMO lasts for five days. Both the type and the difficulty of the problems match those of IMO. Similarly, they solve three problems every day in four and half hours. From CMO, about 20 to 30 students are selected to form a national training team. The training lasts for two weeks in March every year. After six to eight tests, plus two qualifying examinations, six students are finally selected to form the national team, to take part in IMO in July that year.

Because of the differences in education, culture and economy of West China in comparison with East China, mathematical competitions in the west did not develop as fast as in the east. In order to promote the activity of mathematical competition there, China Mathematical Olympiad Committee conducted the China Western Mathematical Olympiad from 2001. The top two winners will be admitted to the national training team. Through the China Western Mathematical Olympiad, there have been two students who entered the national team and received Gold Medals at IMO.

Since 1986, the china team has never had a female student. In order to encourage more female students to participate in the mathematical competition, starting from 2002, China Mathematical Olympiad Committee conducted the China Girls' mathematical Olympiad. Again, the top two winners will be admitted directly into the national training team.

The authors of this book are coaches of the China national team. They are Xiong Bin, Li Shenghong, Chen Yonggao, Leng Gangsong, Wang Jianwei, Li Weigu, Zhu Huawei, Feng Zhigang, Wang Haiming, Xu Wenbin, Tao Pingshen, and Zheng Chongyi. Those who took part in the translation work are Xiong Bin, Feng Zhigang, Ma Guoxuan, Lin Lei, Wang Shanping, Zheng Chongyi, and Hao Lili. We are grateful to Qiu Zhonghu, Wang Jie, Wu Jianping, and Pan Chengbiao for their guidance and assistance to authors. We are grateful to Ni Ming and Xu Jin of East China Normal University Press. Their effort has helped make our job easier. We are also grateful to Zhang Ji of World Scientific Publishing for her hard work leading to the final publication of the book.

Authors
March 2007

Introduction

Early days

The International Mathematical Olympiad (IMO), founded in 1959, is one of the most competitive and highly intellectual activities in the world for high school students.

Even before IMO, there were already many countries which had mathematics competition. They were mainly the countries in Eastern Europe and in Asia. In addition to the popularization of mathematics and the convergence in educational systems among different countries, the success of mathematical competitions at the national level provided a foundation for the setting-up of IMO. The countries that asserted great influence are Hungary, the former Soviet Union and the United States. Here is a brief review of the IMO and mathematical competition in China.

In 1894, the Department of Education in Hungary passed a motion and decided to conduct a mathematical competition for the secondary schools. The well-known scientist, *J. von Etövös*, was the Minister of Education at that time. His support in the event had made it a success and thus it was well publicized. In addition, the success of his son, *R. von Etövös*, who was also a physicist, in proving the principle of equivalence of the general theory of relativity by *A. Einstein* through experiment, had brought Hungary to the world stage in science. Thereafter, the prize for mathematics competition in Hungary was named "*Etövös* prize". This was the first formally organized mathematical competition in the world. In what follows,

Hungary had indeed produced a lot of well-known scientists including *L. Fejér*, *G. Szegö*, *T. Radó*, *A. Haar* and *M. Riesz* (in real analysis), *D. König* (in combinatorics), *T. von Kármán* (in aerodynamics), and *J. C. Harsanyi* (in game theory, who had also won the Nobel Prize for Economics in 1994). They all were the winners of Hungary mathematical competition. The top scientific genius of Hungary, *J. von Neumann*, was one of the leading mathematicians in the 20th century. *Neumann* was overseas while the competition took place. Later he did it himself and it took him half an hour to complete. Another mathematician worth mentioning is the highly productive number theorist *P. Erdös*. He was a pupil of *Fejér* and also a winner of the Wolf Prize. *Erdös* was very passionate about mathematical competition and setting competition questions. His contribution to discrete mathematics was unique and greatly significant. The rapid progress and development of discrete mathematics over the subsequent decades had indirectly influenced the types of questions set in IMO. An internationally recognized prize named after Erdös was to honour those who had contributed to the education of mathematical competition. Professor *Qiu Zonghu* from China had won the prize in 1993.

In 1934, *B. Delone*, a famous mathematician, conducted a mathematical competition for high school students in Leningrad (now St. Petersburg). In 1935, Moscow also started organizing such event. Other than being interrupted during the World War II, these events had been carried on until today. As for the Russian Mathematical Competition (later renamed as the Soviet Mathematical Competition), it was not started until 1961. Thus, the former Soviet Union and Russia became the leading powers of Mathematical Olympiad. A lot of grandmasters in mathematics including *A. N. Kolmogorov* were all very enthusiastic about the mathematical competition. They would personally involve in setting the questions for the competition. The former Soviet Union even called it the Mathematical Olympiad, believing that mathematics is the "gymnastics of thinking". These points of view gave a great impact on the educational community. The winner of the Fields Medal in 1998, *M. Kontsevich*, was once the first runner-up of the Russian Mathematical Competition. *G. Kasparov*, the international chess grandmaster, was once the second runner-up. *Grigori Perelman*, the winner of the Fields Medal in 2006, who solved the Poincaré's Conjecture, was a gold medalist of IMO in 1982.

In the United States of America, due to the active promotion by the renowned mathematician *Birkhoff* and his son, together with *G. Pólya*, the Putnam mathematics competition was organized in 1938 for junior undergraduates. Many of the questions were within the scope of high school students. The top five contestants of the Putnam mathematical competition would be entitled to the membership of Putnam. Many of these were eventually outstanding mathematicians. There were *R. Feynman* (winner of the Nobel Prize for Physics, 1965), *K. Wilson* (winner of the Nobel Prize for Physics, 1982), *J. Milnor* (winner of the Fields Medal, 1962), *D. Mumford* (winner of the Fields Medal, 1974), *D. Quillen* (winner of the Fields Medal, 1978), et al.

Since 1972, in order to prepare for the IMO, the United States of American Mathematical Olympiad (USAMO) was organized. The standard of questions posed was very high, parallel to that of the Winter Camp in China. Prior to this, the United States had organized American High School Mathematics Examination (AHSME) for the high school students since 1950. This was at the junior level yet the most popular mathematics competition in America. Originally, it was planned to select about 100 contestants from AHSME to participate in USAMO. However, due to the discrepancy in the level of difficulty between the two competitions and other restrictions, from 1983 onwards, an intermediate level of competition, namely, American Invitational Mathematics Examination (AIME), was introduced. Henceforth both AHSME and AIME became internationally well-known. A few cities in China had participated in the competition and the results were encouraging.

The members of the national team who were selected from USAMO would undergo training at the West Point Military Academy, and would meet the President at the White House together with their parents. Similarly as in the former Soviet Union, the Mathematical Olympiad education was widely recognized in America. The book "How to Solve it" written by *George Polya* along with many other titles had been translated into many different languages. *George Polya* provided a whole series of general heuristics for solving problems of all kinds. His influence in the educational community in China should not be underestimated.

International Mathematical Olympiad

In 1956, the East European countries and the Soviet Union took the initiative to organize the IMO formally. The first International Mathematical Olympiad (IMO) was held in Brasov, Romania, in 1959. At the time, there were only seven participating countries, namely, Romania, Bulgaria, Poland, Hungary, Czechoslovakia, East Germany and the Soviet Union. Subsequently, the United States of America, United Kingdom, France, Germany and also other countries including those from Asia joined. Today, the IMO had managed to reach almost all the developed and developing countries. Except in the year 1980 due to financial difficulties faced by the host country, Mongolia, there were already 47 Olympiads held and 90 countries participating.

The mathematical topics in the IMO include number theory, polynomials, functional equations, inequalities, graph theory, complex numbers, combinatorics, geometry and game theory. These areas had provided guidance for setting questions for the competitions. Other than the first few Olympiads, each IMO is normally held in mid-July every year and the test paper consists of 6 questions in all. The actual competition lasts for 2 days for a total of 9 hours where participants are required to complete 3 questions each day. Each question is 7 marks which total up to 42 marks. The full score for a team is 252 marks. About half of the participants will be awarded a medal, where 1/12 will be awarded a gold medal. The numbers of gold, silver and bronze medals awarded are in the ratio of 1:2:3 approximately. In the case when a participant provides a better solution than the official answer, a special award is given.

Each participating country will take turn to host the IMO. The cost is borne by the host country. China had successfully hosted the 31st IMO in Beijing in 1990. The event had made a great impact on the mathematical community in China. According to the rules and regulations of the IMO, all participating countries are required to send a delegation consisting of a leader, a deputy leader and 6 contestants. The problems are contributed by the participating countries and are later selected carefully by the host country for submission to the international jury set up by the host country. Eventually, only 6 problems will be accepted for use in the competition. The host country does not provide any question. The short-listed problems are subsequently translated, if necessary, in English, French, German, Russian and other working languages. After that, the team leaders will translate the problems into their own languages.

The answer scripts of each participating team will be marked by the team leader and the deputy leader. The team leader will later present the scripts of their contestants to the coordinators for assessment. If there is any dispute, the matter will be settled by the jury. The jury is formed by the various team leaders and an appointed chairman by the host country. The jury is responsible for deciding the final 6 problems for the competition. Their duties also include finalizing the marking standard, ensuring the accuracy of the translation of the problems, standardizing replies to written queries raised by participants during the competition, synchronizing differences in marking between the leaders and the coordinators and also deciding on the cut-off points for the medals depending on the contestants' results as the difficulties of problems each year are different.

China had participated informally in the 26th IMO in 1985. Only two students were sent. Starting from 1986, except in 1998 when the IMO was held in Taiwan, China had always sent 6 official contestants to the IMO. Today, the Chinese contestants not only performed outstandingly in the IMO, but also in the International Physics, Chemistry, Informatics, and Biology Olympiads. So far, no other countries have overtaken China in the number of gold and silver medals received. This can be regarded as an indication that China pays great attention to the training of basic skills in mathematics and science education.

Winners of the IMO

Among all the IMO medalists, there were many of them who eventually became great mathematicians. Some of them were also awarded the Fields Medal, Wolf Prize or Nevanlinna Prize (a prominent mathematics prize for computing and informatics). In what follows, we name some of the winners.

G. Margulis, a silver medalist of IMO in 1959, was awarded the Fields Medal in 1978. *L. Lovasz*, who won the Wolf Prize in 1999, was awarded the Special Award in IMO consecutively in 1965 and 1966. *V. Drinfeld*, a gold medalist of IMO in 1969, was awarded the Fields Medal in 1990. *J. -C. Yoccoz* and *T. Gowers*, who were both awarded the Fields Medal in 1998, were gold medalists in IMO in 1974 and 1981 respectively. A silver medalist of IMO in 1985, *L. Lafforgue*, won the Fields Medal in 2002. A gold medalist of IMO in 1982, *Grigori Perelman* from Russia, was awarded the Fields Medal in 2006 for solving the final step of the Poincaré conjecture. In 1986, 1987, and 1988, *Terence Tao* won a bronze, silver, and gold medal respectively. He was the youngest participant to date in the IMO, first competing at the age of ten. He was also awarded the Fields Medal in 2006.

A silver medalist of IMO in 1977, *P. Shor*, was awarded the Nevanlinna Prize. A gold medalist of IMO in 1979, *A. Razborov*, was awarded the Nevanlinna Prize. Another gold medalist of IMO in 1986, *S. Smirnov*, was awarded the Clay Research Award. *V. Lafforgue*, a gold medalist of IMO in 1990, was awarded the European Mathematical Society prize. He is *L. Laforgue*'s younger brother.

Also, a famous mathematician in number theory, *N. Elkis*, who is also a foundation professor at Havard University, was awarded a gold medal of IMO in 1981. Other winners include *P. Kronheimer* awarded a silver medal in 1981 and *R. Taylor* a contestant of IMO in 1980.

Mathematical competitions in China

Due to various reasons, mathematical competitions in China started relatively late but is progressing vigorously.

"We are going to have our own mathematical competition too!" said *Hua Luogeng*. *Hua* is a house-hold name in China. The first mathematical competition was held concurrently in Beijing, Tianjing, Shanghai and Wuhan in 1956. Due to the political situation at the time, this event was interrupted a few times. Until 1962, when the political environment started to improve, Beijing and other cities started organizing the competition though not regularly. In the era of cultural revolution, the whole educational system in China was in chaos. The mathematical competition came to a complete halt. In contrast, the mathematical competition in the former Soviet Union was still on-going during the war and at a time under the difficult political situation. The competitions in Moscow were interrupted only 3 times between 1942 and 1944. It was indeed commendable.

In 1978, it was the spring of science. *Hua Luogeng* conducted the Middle School Mathematical Competition for 8 provinces in China. The mathematical competition in China was then making a fresh start and embarked on a road of rapid development. *Hua* passed away in

1985. In commemorating him, a competition named *Hua Luogeng Gold Cup* was set up in 1986 for the junior middle school students and it had a great impact.

The mathematical competitions in China before 1980 can be considered as the initial period. The problems set were within the scope of middle school textbooks. After 1980, the competitions were gradually moving towards the senior middle school level. In 1981, the Chinese Mathematical Society decided to conduct the China Mathematical Competition, a national event for high schools.

In 1981, the United States of America, the host country of IMO, issued an invitation to China to participate in the event. Only in 1985, China sent two contestants to participate informally in the IMO. The results were not encouraging. In view of this, another activity called the Winter Camp was conducted after the China Mathematical Competition. The Winter Camp was later renamed as the China Mathematical Olympiad or CMO. The winning team would be awarded the *Chern Shiing-Shen* Cup. Based on the outcome at the Winter Camp, a selection would be made to form the 6-member national team for IMO. From 1986 onwards, other than the year when IMO was organized in Taiwan, China had been sending a 6-member team to IMO every year. China is normally awarded the champion or first runner-up except on three occasions when the results were lacking. Up to 2006, China had been awarded the overall team champion for 13 times.

In 1990, China had successfully hosted the 31st IMO. It showed that the standard of mathematical competition in China has leveled that of other leading countries. First, the fact that China achieves the highest marks at the 31st IMO for the team is an evidence of the effectiveness of the pyramid approach in selecting the contestants in China. Secondly, the Chinese mathematicians had simplified and modified over 100 problems and submitted them to the team leaders of the 35 countries for their perusal. Eventually, 28 problems were recommended. At the end, 5 problems were chosen (IMO requires 6 problems). This is another evidence to show that China has achieved the highest quality in setting problems. Thirdly, the answer scripts of the participants were marked by the various team leaders and assessed by the coordinators who were nominated by the host countries. China had formed a group 50 mathematicians to serve as coordinators who would ensure the high accuracy and fairness in marking. The marking process was completed half a day earlier than it was scheduled. Fourthly, that was the first ever IMO organized in Asia. The outstanding performance by China had encouraged the other developing countries, especially those in Asia. The organizing and coordinating work of the IMO by the host country was also reasonably good.

In China, the outstanding performance in mathematical competition is a result of many contributions from all the quarters of mathematical community. There are the older generation of mathematicians, middle-aged mathematicians and also the middle and elementary school teachers. There is one person who deserves a special mention and he is *Hua Luogeng*. He initiated and promoted the mathematical competition. He is also the author of the following books: Beyond *Yang hui*'s Triangle, Beyond the *pi* of *Zu Chongzhi*, Beyond the Magic Computation of *Sun-zi*, Mathematical Induction, and Mathematical Problems of Bee Hive. These books were derived from mathematics competitions. When China resumed mathematical competition in 1978, he participated in setting problems and giving critique to solutions of the problems. Other outstanding books derived from the Chinese mathematics competitions are: Symmetry by *Duan Xuefu*, Lattice and Area by *He Sihe*, One Stroke Drawing and Postman Problem by *Jiang Boju*.

After 1980, the younger mathematicians in China had taken over from the older generation of mathematicians in running the mathematical competition. They worked and strived hard to bring the level of mathematical competition in China to a new height. *Qiu Zonghu* is one such outstanding representative. From the training of contestants and leading the team 3 times to IMO to the organizing of the 31th IMO in China, he had contributed prominently and was awarded the *P. Erdös* prize.

Preparation for IMO

Currently, the selection process of participants for IMO in China is as follows.

First, the China Mathematical Competition, a national competition for high Schools, is organized on the second Sunday in October every year. The objectives are: to increase the interest of students in learning mathematics, to promote the development of co-curricular activities in mathematics, to help improve the teaching of mathematics in high schools, to discover and cultivate the talents and also to prepare for the IMO. This happens since 1981. Currently there are about 200 000 participants taking part.

Through the China Mathematical Competition, around 120 of students are selected to take part in the China Mathematical Olympiad or CMO, that is, the Winter Camp. The CMO lasts for 5 days and is held in January every year. The types and difficulties of the problems in CMO are very much similar to the IMO. There are also 3 problems to be completed within four and half hours each day. However, the score for each problem is 21 marks which add up to 126 marks in total. Starting from 1990, the Winter Camp instituted the *Chern Shiing-Shen* Cup for team championship. In 1991, the Winter Camp was officially renamed as the China Mathematical Olympiad (CMO). It is similar to the highest national mathematical competition in the former Soviet Union and the United States.

The CMO awards the first, second and third prizes. Among the participants of CMO, about 20 to 30 students are selected to participate in the training for IMO. The training takes place in March every year. After 6 to 8 tests and another 2 rounds of qualifying examinations, only 6 contestants are short-listed to form the China IMO national team to take part in the IMO in July.

Besides the China Mathematical Competition (for high schools), the Junior Middle School Mathematical Competition is also developing well. Starting from 1984, the competition is organized in April every year by the Popularization Committee of the Chinese Mathematical Society. The various provinces, cities and autonomous regions would rotate to host the event. Another mathematical competition for the junior middle schools is also conducted in April every year by the Middle School Mathematics Education Society of the Chinese Educational Society since 1998 till now.

The *Hua Luogeng* Gold Cup, a competition by invitation, had also been successfully conducted since 1986. The participating students comprise elementary six and junior middle one students. The format of the competition consists of a preliminary round, semifinals in various provinces, cities and autonomous regions, then the finals.

Mathematical competition in China provides a platform for students to showcase their talents in mathematics. It encourages learning of mathematics among students. It helps identify talented students and to provide them with differentiated learning opportunity. It develops co-curricular activities in mathematics. Finally, it brings about changes in the teaching of mathematics.

Contents

China Mathematical Competition

The China Mathematical Competition is organized in October every year. The Popularization Committee of the Chinese Mathematical Society and the local Mathematical Society are responsible for the assignments of the competition problems.

The test paper consists of 6 choices, 6 blanks and 3 questions to be solved with complete process. The full score is 150 marks. Besides, 3 questions are used in the Extra Test, with 50 marks each. The participants with high total marks in the China Mathematical Competition plus the Extra Test are awarded the first prize (1 000 participants around China), and they will be admitted into the university directly.

The participants with excellent marks are selected to take part in the China Mathematical Olympiad the next year. Thus, for Chinese high school students, the China Mathematical Olympiad is the first step to IMO.

2002 (Jilin)

Popularization Committee of CMS and Jilin Mathematical Society were responsible for the assignment of the competition problems in the first and second rounds of the contests.

Part I　**Multiple-choice Questions (Questions 1 to 6 carry 6 marks each.)**

1　The interval on which the function $f(x) = \log_{\frac{1}{2}}(x^2 - 2x - 3)$ is monotone increasing is (　　).

　　(A) $(-\infty, -1)$　　　　　　　(B) $(-\infty, 1)$

　　(C) $(1, +\infty)$　　　　　　　(D) $(3, +\infty)$

Solution　First, we will find the domain of $f(x)$. From $x^2 - 2x - 3 > 0$, we obtain $x < -1$ or $x > 3$. So the domain of definition for $f(x)$ is $(-\infty, -1) \cup (3, +\infty)$. But $u = x^2 - 2x - 3 = (x-1)^2 - 4$ is monotone decreasing on $(-\infty, -1)$, and monotone increasing on $(3, +\infty)$. So $f(x) = \log_{\frac{1}{2}}(x^2 - 2x - 3)$ is monotone increasing on $(-\infty, -1)$, and monotone decreasing on $(3, +\infty)$. Answer: A.

2　If real numbers x and y satisfy $(x+5)^2 + (y-12)^2 = 14^2$, then the minimum value of $x^2 + y^2$ is (　　).

　　(A) 2　　　　　(B) 1　　　　(C) $\sqrt{3}$　　　　(D) $\sqrt{2}$

Solution　Let $x + 5 = 14\cos\theta$ and $y - 12 = 14\sin\theta$, for $\theta \in [0, 2\pi)$. Hence

$$x^2 + y^2 = (14\cos\theta - 5)^2 + (14\sin\theta + 12)^2$$

$$= 14^2 + 5^2 + 12^2 - 140\cos\theta + 336\sin\theta$$

$$= 365 + 28(12\sin\theta - 5\cos\theta)$$

$$=365+28 \times 13\sin(\theta-\varphi)$$

$$=365+364\sin(\theta-\varphi),$$

where $\tan \varphi = \dfrac{5}{12}$.

So x^2+y^2 has the minimum value 1, when $\theta = \dfrac{3\pi}{2}+\arctan\dfrac{5}{12}$, i.e.

$x = \dfrac{5}{13}$ and $y = -\dfrac{12}{13}$. Answer: B.

Remark A geometric significance of this problem is: $(x+5)^2 + (y-12)^2 = 14^2$ is a circle with $C(-5, 12)$ as center and 14 as radius. We can find a point P on the circumference of this circle such that $|PO|^2$ is minimal, where O is the origin of the coordinate system. We join CO and extend it to intersect the circumference of the circle at P. Then it follows that $|PO|^2$ is the minimum value of $x^2 +y^2$.

The function $f(x)= \dfrac{x}{1-2^x} - \dfrac{x}{2}$ is (   ).

(A) an even but not odd function

(B) an odd but not even function

(C) a both even and odd function

(D) a neither even nor odd function

Solution It is easy to see that the domain of $f(x)$ is $(-\infty, 0) \cup (0, +\infty)$. When $x \in (-\infty, 0) \cup (0, +\infty)$, we have

$$f(-x) = \frac{-x}{1-2^{-x}} - \frac{-x}{2}$$

$$= \frac{-x \cdot 2^x}{2^x -1} + \frac{x}{2}$$

$$= \frac{x}{1-2^x} - x + \frac{x}{2}$$

$$= \frac{x}{1-2^x} - \frac{x}{2} = f(x).$$

Therefore, $f(x)$ is an even function, and obviously not an odd function. Answer: A.

⚫ The straight line $\frac{x}{4} + \frac{y}{3} = 1$ intersects the ellipse $\frac{x^2}{16} + \frac{y^2}{9} = 1$ at two points A and B. There is a point P on this ellipse such that the area of $\triangle PAB$ is equal to 3. There is/are (   ) such point/points P.

   (A) 1   (B) 2   (C) 3   (D) 4

Solution   Suppose that there is a point $P(4\cos\alpha, 3\sin\alpha)$ on the ellipse. When P and the origin O are not on the same side of AB, the distance from P to AB is

$$\frac{3(4\cos\alpha) + 4(3\sin\alpha) - 12}{5}$$

$$= \frac{12}{5}(\cos\alpha + \sin\alpha - 1)$$

$$\leqslant \frac{12}{5}(\sqrt{2} - 1)$$

$$< \frac{6}{5}.$$

But $AB = 5$, so $\triangle PAB < \frac{1}{2} \times 5 \times \frac{6}{5} = 3$.

   Therefore, when the area of $\triangle PAB$ is equal to 3, points P and O are on the same side of AB. There are two such points P. Answer: B.

⚫ It is given that there are two sets of real numbers $A = \{a_1, a_2, \cdots, a_{100}\}$ and $B = \{b_1, b_2, \cdots, b_{50}\}$. If there is a mapping f from A to B such that every element in B has an inverse image and

$$f(a_1) \leqslant f(a_2) \leqslant \cdots \leqslant f(a_{100}),$$

then the number of such mappings is (   ).

   (A) C_{100}^{50} (B) C_{99}^{50} (C) C_{100}^{49} (D) C_{99}^{49}

Solution   We might as well suppose $b_1 < b_2 < \cdots < b_{50}$, and divide elements a_1, a_2, $\cdots$, a_{100} in A into 50 nonempty groups according to their order. Define a mapping $f: A \to B$, so that the images of all the elements in the i-th group are b_i $(i = 1, 2, \cdots, 50)$ under the mapping. Obviously, f satisfies the requirements given in the problem. Furthermore, there is a one-to-one correspondence between all groups so divided and the mappings satisfying the condition. So the number of mappings f satisfying the requirements is equal to the number of ways dividing A into 50 groups according to the order of the subscripts. The number of ways dividing A is C_{99}^{49}. Then there are, in all, C_{99}^{49} such mappings. Answer: D.

Remark Since $C_{99}^{50} = C_{99}^{49}$, Answer B is also true in this problem. This may be an oversight when the problem was set.

A region is enclosed by the curves $x^2 = 4y$, $x^2 = -4y$, $x = 4$ and $x = -4$. V_1 is the volume of the solid obtained by rotating the above region round the y-axis. Another region consists of points (x, y) satisfying $x^2 + y^2 \leqslant 16$, $x^2 + (y-2)^2 \geqslant 4$ and $x^2 + (y+2)^2 \geqslant 4$. V_2 is the volume of the solid obtained by rotating this region round the y-axis. Then ( ).

(A) $V_1 = \dfrac{1}{2}V_2$ (B) $V_1 = \dfrac{2}{3}V_2$

(C) $V_1 = V_2$ (D) $V_1 = 2V_2$

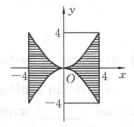
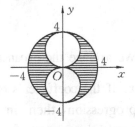

Solution As shown in the diagram, two solids of rotation obtained by rotating respectively two regions round the y-axis lie between two parallel planes, which are 8 units apart. We cut two solids of rotation by any plane which is perpendicular to the y-axis. Suppose the distance from the plane to the origin is $|y|$ ($\leqslant 4$). Then the two sectional areas are

$$S_1 = \pi(4^2 - 4|y|),$$

and $\qquad S_2 = \pi(4^2 - y^2) - \pi[4 - (2 - |y|)^2] = \pi(4^2 - 4|y|).$

So $\qquad\qquad\qquad\qquad S_1 = S_2.$

From the Zugen Principle (or Cavalieri Principle), we know that the volumes of the two geometric solids are equal, and that is, $V_1 = V_2$. Answer: C.

Part II   Short-answer Questions (Questions 7 to 12 carry 6 marks each.)

It is given that complex numbers z_1 and z_2 satisfy $|z_1| = 2$ and $|z_2| = 3$. If the included angle of their corresponding vectors is $60°$, then $\left| \dfrac{z_1 + z_2}{z_1 - z_2} \right| = $ _____ .

Solution   By the cosine rule, we obtain

$$|z_1 + z_2| = \sqrt{|z_1|^2 + |z_2|^2 - 2|z_1||z_2|\cos 120°} = \sqrt{19},$$

and $\qquad |z_1 - z_2| = \sqrt{|z_1|^2 + |z_2|^2 - 2|z_1||z_2|\cos 60°} = \sqrt{7}.$

Therefore, $\qquad\qquad \left| \dfrac{z_1 + z_2}{z_1 - z_2} \right| = \dfrac{\sqrt{19}}{\sqrt{7}} = \dfrac{\sqrt{133}}{7}.$

We arrange the expansion of $\left(\sqrt{x} + \dfrac{1}{2\sqrt[4]{x}} \right)^n$ in decreasing powers of x. If the coefficients of the first three terms form an arithmetic progression, then, in the expansion, there are _____ terms of x with integer power.

Solution   The coefficients of the first three terms are 1, $\dfrac{1}{2}C_n^1$ and $\dfrac{1}{2^2}C_n^2$. By the assumption, we have

$$2 \cdot \frac{1}{2} C_n^1 = 1 + \frac{1}{2^2} C_n^2.$$

That is

$$n = 1 + \frac{1}{8} n(n-1).$$

Solving for n, we obtain $n = 8$ and $n = 1$ (not admissible).

When $n = 8$, $T_{r+1} = C_8^r \left(\frac{1}{2}\right)^r x^{\frac{16-3r}{4}}$, $r = 0, 1, 2, \cdots, 8$. But r must satisfy $4 \mid 16 - 3r$, so r can only take 0, 4 and 8. Therefore, there are 3 terms in the expansion of x with integer power.

9 As shown in the diagram, points P_1, P_2, $\cdots$, P_{10} are either the vertices or the midpoints of the edges of a tetrahedron respectively. Then there are _____ groups of four points $(P_1,$ $P_i, P_j, P_k)$ $(1 < i < j < k \leqslant 10)$ on the same plane.

Solution   On each lateral face of the tetrahedron other than point P_1 there are five points. Take any three out of the five points and add point P_1. These four points lie in the same plane. There are C_5^3 such groups for each lateral face. Hence there are $3C_5^3$ groups in all for three lateral faces.

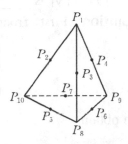

Furthermore, there are three points on each edge containing point P_1. We add a midpoint taking from the edge on the base, which is not on the same plane with the edge above. Now, we obtain another group consisting of four points which also are on the same plane. There are 3 groups like this.

Consequently, there are $3C_5^3 + 3 = 33$ groups of four points on the same plane.

10 It is given that $f(x)$ is a function defined on $\mathbf{R}$, satisfying $f(1) = 1$, and for any $x \in \mathbf{R}$,

$$f(x+5) \geqslant f(x) + 5,$$

and $$f(x+1) \leqslant f(x)+1.$$

If $g(x) = f(x)+1-x$, then $g(2\,002) = $ _____ .

Solution   We determine $f(2\,002)$ first. From the conditions given, we have

$$f(x) +5 \leqslant f(x+5) \leqslant f(x+4)+1$$
$$\leqslant f(x+3)+2 \leqslant f(x+2)+3$$
$$\leqslant f(x+1)+4 \leqslant f(x)+5.$$

Thus the equality holds for all. So we have $f(x+1) = f(x)+1$.

Hence, from $f(1) = 1$, we get $f(2) = 2$, $f(3) = 3$, $\cdots$, $f(2\,002) = 2\,002$. Therefore, $g(2\,002) = f(2\,002)+1-2\,002 = 1$.

⬤ If $\log_4(x+2y) + \log_4(x-2y) = 1$, then the minimum value of $|x|-|y|$ is _____ .

Solution   First, from

$$\begin{cases} x+2y > 0, \\ x-2y > 0, \\ (x+2y)(x-2y) = 4, \end{cases}$$

we obtain

$$\begin{cases} x > 2|y| \geqslant 0, \\ x^2 - 4y^2 = 4. \end{cases}$$

By the symmetry, there is no loss of generality in considering only the case when $y \geqslant 0$. In view of $x > 0$, we need to find the minimum value of $x - y$ only.

Setting $u = x-y$, and substituting it into $x^2 - 4y^2 = 4$, we obtain

$$3y^2 - 2uy + (4 - u^2) = 0. \qquad (*)$$

Equation $(*)$ with respect to y has real solutions. So we have

$$\Delta = 4u^2 - 12(4 - u^2) \geqslant 0.$$

Thereby $$u \geqslant \sqrt{3}.$$

In addition, when $x = \dfrac{4}{3}\sqrt{3}$ and $y = \dfrac{\sqrt{3}}{3}$, we have $u = \sqrt{3}$.

Therefore, the minimum value of $|x| - |y|$ is $\sqrt{3}$.

12 If the inequality

$$\sin^2 x + a\cos x + a^2 \geqslant 1 + \cos x$$

holds for any $x \in \mathbf{R}$, the range of values for negative a is

_____.

Solution $a + a^2 \geqslant 2$ when $x = 0$. So $a \leqslant -2$ (because $a < 0$). When $a \leqslant -2$, we have

$$a^2 + a\cos x \geqslant a^2 + a \geqslant 2 \geqslant \cos^2 x + \cos x$$
$$= 1 + \cos x - \sin^2 x,$$

that is, $\qquad \sin^2 x + a\cos x + a^2 \geqslant 1 + \cos x.$

Hence, the range of values for negative a is $a \leqslant -2$.

Part III   Word Problems (Questions 13 to 15 carry 20 marks each.)

13 Given $A(0, 2)$ and two points B and C on the parabola $y^2 = x + 4$ such that $AB \perp BC$, determine the range for the y-coordinate of point C.

Solution Suppose that $(y_1^2 - 4, y_1)$ is the coordinates of point B and $(y^2 - 4, y)$ of point C. Obviously, $y_1^2 - 4 \neq 0$, so $k_{AB} = \dfrac{y_1 - 2}{y_1^2 - 4} = \dfrac{1}{y_1 + 2}.$

Since $AB \perp BC$, so $k_{BC} = -(y_1 - 2)$. Thus

$$y - y_1 = -(y_1 + 2)[y^2 - 4 - (y_1^2 - 4)].$$

Noting $y \neq y_1$, we obtain

$$(2 + y_1)(y + y_1) + 1 = 0,$$

and that is $\qquad y_1^2 + (2 + y)y_1 + (2y + 1) = 0.$

From $\Delta \geqslant 0$, we obtain $y \leqslant 0$ or $y \geqslant 4$.

When $y = 0$, the coordinates of B are $(-3, -1)$ and when $y = 4$, they are $(5, -3)$. They both satisfy the conditions given by the problem. So, the range of values for the y-coordinate of point C is $y \leqslant 0$ or $y \geqslant 4$.

⑤ As shown in the diagram, there is a sequence of the curves P_0, P_1, P_2, $\cdots$. It is known that the region enclosed by P_0 has area 1 and P_0 is an equilateral triangle. We obtain P_{k+1} from P_k by operating as follows: Trisecting every side of P_k, then we construct an equilateral triangle outwardly on every side of P_k sitting on the middle segment of the side and finally remove this middle segment $(k = 0, 1, 2, \cdots)$. Write S_n as the area of the region enclosed by P_n.

(1) Find a formula for the general term of the sequence of numbers $\{S_n\}$;

(2) Find $\lim\limits_{n \to \infty} S_n$.

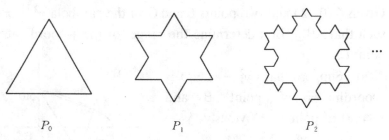

P_0 P_1 P_2

Solution (1) We perform the operation on P_0. It is easy to see that each side of P_0 becomes 4 sides of P_1. So the number of sides of P_1 is $3 \cdot 4$. In the same way, we operate on P_1. Each side of P_1 becomes 4 sides of P_2. So the number of sides of P_2 is $3 \cdot 4^2$. Consequently, it is not difficult to get that the number of sides of P_n is $3 \cdot 4^n$.

It is known that the area of P_0 is $S_0 = 1$. Comparing P_1 with P_0, it is easy to see that we add to P_1 a smaller equilateral triangle with area $\dfrac{1}{3^2}$ on each side of P_0. Since P_0 has 3 sides, so $S_1 = S_0 + 3 \cdot \dfrac{1}{3^2} =$

$1 + \frac{1}{3}$. Again, comparing P_2 with P_1, we see that P_2 has an additional smaller equilateral triangle with area $\frac{1}{3^2} \cdot \frac{1}{3^2}$ on each side of P_1, and P_1 has $3 \cdot 4$ sides. So that

$$S_2 = S_1 + 3 \cdot 4 \cdot \frac{1}{3^4} = 1 + \frac{1}{3} + \frac{4}{3^3}.$$

Similarly, we have

$$S_3 = S_2 + 3 \cdot 4 \cdot \frac{1}{3^6} = 1 + \frac{1}{3} + \frac{4}{3^3} + \frac{4^2}{3^5}.$$

Hence, we have

$$S_n = 1 + \frac{1}{3} + \frac{4}{3^3} + \frac{4^2}{3^5} + \cdots + \frac{4^{n-1}}{3^{2n-1}}$$

$$= 1 + \sum_{k=1}^{n} \frac{4^{k-1}}{3^{2k-1}} = 1 + \frac{3}{4} \sum_{k=1}^{n} \left(\frac{4}{9} \right)^k$$

$$= 1 + \frac{3}{4} \cdot \frac{\frac{4}{9} \cdot \left[1 - \left(\frac{4}{9} \right)^n \right]}{1 - \frac{4}{9}}$$

$$= 1 + \frac{3}{5} \left[1 - \left(\frac{4}{9} \right)^n \right]$$

$$= \frac{8}{5} - \frac{3}{5} \cdot \left(\frac{4}{9} \right)^n. \tag{$*$}$$

We will prove ($*$) by mathematical induction as follows:

When $n = 1$, it is known that ($*$) holds from above.

Suppose, when $n = k$, we have $S_k = \frac{8}{5} - \frac{3}{5} \cdot \left(\frac{4}{9} \right)^k$.

When $n = k + 1$, it is easy to see that, after $k + 1$ times of operations, by comparing P_{k+1} with P_k, we have added to P_{k+1} a smaller equilateral triangle with area $\frac{1}{3^{2(k+1)}}$ on each side of P_k and

P_k has $3 \cdot 4^k$ sides. So we get

$$S_{k+1} = S_k + 3 \cdot 4^k \cdot \frac{1}{3^{2(k+1)}}$$

$$= S_k + \frac{4^k}{3^{2k+1}} = \frac{8}{5} - \frac{3}{5} \cdot \left(\frac{4}{9}\right)^{k+1}.$$

By mathematical induction, (*) is proved.

(2) From (1), we have $S_n = \frac{8}{5} - \frac{3}{5} \cdot \left(\frac{4}{9}\right)^n$.

Therefore, $\lim\limits_{n \to \infty} S_n = \lim\limits_{n \to \infty}\left[\frac{8}{5} - \frac{3}{5} \cdot \left(\frac{4}{9}\right)^n\right] = \frac{8}{5}$.

⑬ Suppose a quadratic function $f(x) = ax^2 + bx + c$ ($a, b, c \in \mathbf{R}$, and $a \neq 0$) satisfies the following conditions:

(1) When $x \in \mathbf{R}$, $f(x-4) = f(2-x)$ and $f(x) \geqslant x$.

(2) When $x \in (0, 2)$, $f(x) \leqslant \left(\frac{x+1}{2}\right)^2$.

(3) The minimum value of $f(x)$ on $\mathbf{R}$ is 0.

Find the maximal m ($m > 1$) such that there exists $t \in \mathbf{R}$, $f(x+t) \leqslant x$ holds so long as $x \in [1, m]$.

Analysis We will determine the analytic expression for $f(x)$ by the known conditions first. Then discuss about m and t, and finally determine the maximal value for m.

Solution Since $f(x-4) = f(2-x)$ for $x \in \mathbf{R}$, it is known that the quadratic function $f(x)$ has $x = -1$ as its axis of symmetry. By condition (3), we know that $f(x)$ opens upward, that is, $a > 0$. Hence

$$f(x) = a(x+1)^2 (a > 0).$$

By condition (1), we get $f(1) \geqslant 1$ and by (2), $f(1) \leqslant \left(\frac{1+1}{2}\right)^2 = 1$. It follows that $f(1) = 1$, i.e. $a(1+1)^2 = 1$. So $a = \frac{1}{4}$.

Thereby, $$f(x) = \frac{1}{4}(x+1)^2.$$

Since the graph of the parabola $f(x) = \frac{1}{4}(x+1)^2$ opens upward, and a graph of $y = f(x+t)$ can be obtained by translating that of $f(x)$ by t units. If we want the graph of $y = f(x+t)$ to lie under the graph of $y = x$ when $x \in [1, m]$, and m to be maximal, then 1 and m should be two roots of an equation with respect to x

$$\frac{1}{4}(x+t+1)^2 = x. \qquad \textcircled{1}$$

Substituting $x = 1$ into $\textcircled{1}$, we get $t = 0$ or $t = -4$.

When $t = 0$, substituting it into $\textcircled{1}$, we get $x_1 = x_2 = 1$ (in contradiction with $m > 1$).

When $t = -4$, substituting it into $\textcircled{1}$, we get $x_1 = 1$, and $x_2 = 9$; and so $m = 9$.

Moreover, when $t = -4$, for any $x \in [1, 9]$, we have always

$$(x-1)(x-9) \leqslant 0$$

$$\Leftrightarrow \frac{1}{4}(x-4+1)^2 \leqslant x,$$

that is $\qquad\qquad\qquad f(x-4) \leqslant x.$

Therefore, the maximum value of m is 9.

2003 (Shaanxi)

Popularization Committee of CMS and Shaanxi Mathematical Society were responsible for the assignment of the competition problems in the first and the second rounds of the contests.

Part I Multiple-choice Questions (Questions 1 to 6 carry 6 marks each.)

A new sequence is obtained from the sequence of the positive integers $\{1, 2, 3, \cdots.\}$ by deleting all the perfect squares. Then the 2 003rd term of the new sequence is (   ).

(A) 2 046   (B) 2 047   (C) 2 048   (D) 2 049

Solution   Since $[\sqrt{2\,046}] = [\sqrt{2\,047}] = [\sqrt{2\,048}] = [\sqrt{2\,049}] = 45$, and $2\,003 + 45 = 2\,048$, Answer: C.

Remark   For any positive whole numbers n and m, satisfying $m^2 < n < (m+1)^2$, we always have $n = a_{n-m}$.

② Suppose a, $b \in \mathbf{R}$, where $ab \neq 0$. Then the graph of the straight line $ax - y + b = 0$ and the conic section $bx^2 + ay^2 = ab$ is (   ).

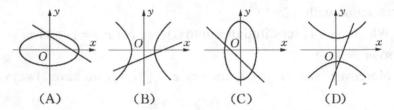

(A)   (B)   (C)   (D)

Solution   In each case, consider b and a, the y-axis intercept and the slope, of the straight line $ax - y + b = 0$. In (A), we have $b > 0$ and $a < 0$, then the conic section must be a hyperbola, and it is impossible. Similarly, (C) is also impossible. We have $a > 0$ and $b < 0$ in both (B) and (D), the conic section is a hyperbola. We have $\dfrac{x^2}{a} - \dfrac{y^2}{-b} = 1$. Answer: B.

③ Let a line with the inclination angle of $60°$ be drawn through the focus F of the parabola $y^2 = 8(x + 2)$. If the two intersection points of the line and the parabola are A and B, and the perpendicular bisector of the chord AB intersects the x-axis at the point P, then the length of the segment PF is (   ).

(A) $\dfrac{16}{3}$ (B) $\dfrac{8}{3}$ (C) $\dfrac{16\sqrt{3}}{3}$ (D) $8\sqrt{3}$

Solution   It follows from the property of the focus of a parabola that $F = (0, 0)$. Then the equation of the straight line through points A and B will be $y = \sqrt{3}x$. Substitute it into the parabola equation, and then obtain

$$3x^2 - 8x - 16 = 0.$$

Let E be the midpoint of the chord AB, then the x-coordinate of E is $\frac{4}{3}$. Then we have $|FE| = \frac{1}{\cos 60°} \times \frac{4}{3} = \frac{8}{3}$, $|PF| = 2|FE| = \frac{16}{3}$.

Answer: A.

Let $x \in \left[-\frac{5\pi}{12}, -\frac{\pi}{3}\right]$. Then the maximum value of

$$y = \tan\left(x + \frac{2\pi}{3}\right) - \tan\left(x + \frac{\pi}{6}\right) + \cos\left(x + \frac{\pi}{6}\right)$$

is (   ).

(A) $\frac{12}{5}\sqrt{2}$ (B) $\frac{11}{6}\sqrt{2}$ (C) $\frac{11}{6}\sqrt{3}$ (D) $\frac{12}{5}\sqrt{3}$

Solution   Let $z = -x - \frac{\pi}{6}$. Then $z \in \left[\frac{\pi}{6}, \frac{\pi}{4}\right]$, and $2z \in \left[\frac{\pi}{3}, \frac{\pi}{2}\right]$.

We have

$$\tan\left(x + \frac{2\pi}{3}\right) = -\cot\left(x + \frac{6}{\pi}\right) = \cot z.$$

Then

$$y = \cot z + \tan z + \cos z = \frac{2}{\sin 2z} + \cos z.$$

Since both $\frac{2}{\sin 2z}$ and $\cos z$ are monotonic decreasing in this case,

so y reaches the maximum at $z = \frac{\pi}{6}$, where $y_{\max} = \frac{2}{\sin\frac{\pi}{3}} + \cos\frac{\pi}{6} =$

$\frac{4}{\sqrt{3}} + \frac{\sqrt{3}}{2} = \frac{11}{6}\sqrt{3}$. Answer: C.

⑤ Suppose x, $y \in (-2, 2)$ and $xy = -1$. Then the minimum value

of $u = \dfrac{4}{4-x^2} + \dfrac{9}{9-y^2}$ is (   ).

(A) $\dfrac{8}{5}$ (B) $\dfrac{24}{11}$ (C) $\dfrac{12}{7}$ (D) $\dfrac{12}{5}$

Solution I   We have

$$u = \frac{4}{4-x^2} + \frac{9x^2}{9x^2-1} = 1 + \frac{35x^2}{-9x^4 + 37x^2 - 4}$$

$$= 1 + \frac{35}{37 - \left(\left(3x - \dfrac{2}{x}\right)^2 + 12\right)}.$$

Since $x \in \left(-2, -\dfrac{1}{2}\right) \cup \left(\dfrac{1}{2}, 2\right)$, so u reaches the minimum value $\dfrac{12}{5}$

when $x = \pm\sqrt{\dfrac{2}{3}}$. Answer: D.

Solution II   It is known from the conditions that $4 - x^2 > 0$ and $9 - y^2 > 0$. Then

$$u \geqslant 2\sqrt{\frac{4}{4-x^2} \cdot \frac{9}{9-y^2}} = \frac{12}{\sqrt{36 - 9x^2 - 4y^2 + (xy)^2}}$$

$$= \frac{12}{\sqrt{37 - 9x^2 - 4y^2}} \geqslant \frac{12}{\sqrt{37 - 2\sqrt{36(xy)^2}}} = \frac{12}{5}.$$

Since u is $\dfrac{12}{5}$ when $x = \sqrt{\dfrac{2}{3}}$ and $y = -\sqrt{\dfrac{3}{2}}$, so u reaches the minimum.

Answer: D.

⑥ Suppose in the tetrahedron $AB\,CD$, $AB = 1$, $CD = \sqrt{3}$, the

distance and angle between the lines AB and CD are 2 and $\dfrac{\pi}{3}$

respectively. Then the volume of the tetrahedron equals (   ).

(A) $\dfrac{\sqrt{3}}{2}$ (B) $\dfrac{1}{2}$ (C) $\dfrac{1}{3}$ (D) $\dfrac{\sqrt{3}}{3}$

Solution   As in the diagram, from point C draw a line CE such that it is equal and parallel to AB. Construct a prism $ABF-ECD$ with $\triangle CDE$ as base and BC as a lateral edge. Denote V_1 as the volume of the tetrahedron and V_2 the volume of the prism, then $V_1 = \frac{1}{3}V_2$.

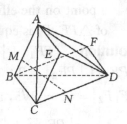

Since $S_{\triangle CDE} = \frac{1}{2}CE \cdot CD \sin \angle ECD$, and the common perpendicular line MN of AB and CD is the height of the prism, then

$$V_2 = \frac{1}{2}MN \cdot CE \cdot CD \sin \angle ECD = \frac{3}{2}. \text{ So } V_1 = \frac{1}{3}V_2 = \frac{1}{2}.$$

Answer: B.

Remark If one is familiar with vector calculus, he will find that the volume of the parallelogram formed by $\overrightarrow{AB}$ and $\overrightarrow{CD}$ is equal to $|\overrightarrow{AB}| \cdot |\overrightarrow{CD}| \cdot \sin 60° = \frac{3}{2}$. Then $S_{\triangle CED} = \frac{1}{2} \times \frac{3}{2} = \frac{3}{4}$. It is easier to find the solution using vectors.

Part II   Short-answer Questions (Questions 7 to 12 carry 6 marks each.)

⬤ The solution set of the inequality $|x|^3 - 2x^2 - 4|x| + 3 < 0$ is

_____.

Solution   Notice that $|x| = 3$ is a root of the equation $|x|^3 - 2x^2 - 4|x| + 3 = 0$. Then the original inequality can be rewritten as $(|x| - 3)(|x|^2 + |x| - 1) < 0$, that is

$$(|x| - 3)\left(|x| - \frac{-1+\sqrt{5}}{2}\right)\left(|x| - \frac{-1-\sqrt{5}}{2}\right) < 0.$$

Since $|x| - \frac{-1-\sqrt{5}}{2} > 0$, then $\frac{-1+\sqrt{5}}{2} < |x| < 3$.

So the solution set is $\left(-3, -\frac{\sqrt{5}-1}{2}\right) \cup \left(\frac{\sqrt{5}-1}{2}, 3\right)$.

⬤ Suppose points F_1, F_2 are the foci of the ellipse $\frac{x^2}{9} + \frac{y^2}{4} = 1$, P is a point on the ellipse, and $|PF_1| : |PF_2| = 2 : 1$. Then the area of $\triangle PF_1F_2$ is equal to _____.

Solution $|PF_1| + |PF_2| = 2a = 6$ by definition of an ellipse. Since $|PF_1| : |PF_2| = 2 : 1$, then $|PF_1| = 4$ and $|PF_2| = 2$. Notice that $|F_1F_2| = 2c = 2\sqrt{5}$, and

$$|PF_1|^2 + |PF_2|^2 = 4^2 + 2^2 = 20 = |F_1F_2|^2.$$

Then $\triangle PF_1F_2$ is a right triangle. So $S_{\triangle PF_1F_2} = \dfrac{1}{2}|PF_1| \cdot |PF_2| = 4$.

⑨ Let $A = \{x \mid x^2 - 4x + 3 < 0, x \in \mathbf{R}\}$, $B = \{x \mid 2^{1-x} + a \leqslant 0,$ $x^2 - 2(a+7)x + 5 \leqslant 0, x \in \mathbf{R}\}$. If $A \subseteq B$, then the range of real number a is _____.

Solution   It is easy to see that $A = (1, 3)$. Now, let

$$f(x) = 2^{1-x} + a, \quad g(x) = x^2 - 2(a+7)x + 5.$$

Then, when $1 < x < 3$, the images of $f(x)$ and $g(x)$ are both below the x-axis since $A \subseteq B$. Using the fact that $f(x)$ is monotone decreasing and $g(x)$ is a quadratic function, we get $A \subseteq B$ if and only if $f(1) \leqslant 0$, $g(1) \leqslant 0$, and $g(3) \leqslant 0$. So the solution is $-4 \leqslant a \leqslant -1$.

⑩ Let a, b, c, d be positive integers and $\log_a b = \dfrac{3}{2}$, $\log_c d = \dfrac{5}{4}$. If $a - c = 9$, then $b - d = $ _____.

Solution   We have $b = a^{\frac{3}{2}}$, $d = c^{\frac{5}{4}}$ from the assumption. We may assume that $a = x^2$, $c = y^4$ with x and y being positive integers, since a, b, c, d are all positive integers. Then $a - c = x^2 - y^4 = (x - y^2)$, $(x + y^2) = 9$. It follows that $(x - y^2, x + y^2) = (1, 9)$. So we obtain the solution $x = 5$, $y = 2$. That is, $b - d = x^3 - y^5 = 125 - 32 = 93$.

⑪ 8 balls of radius 1 are placed in a cylinder in two layers, with each layer containing 4 balls. Each ball is in contact with 2 balls in the same layer, 2 balls in the other layer, one base and the lateral surface of the cylinder. Then the height of the cylinder is

Solution As in the diagram, let A, B, C, D be the centers of the 4 balls in the bottom layer, and A', B', C', D' the centers of the 4 balls in the upper layer. Then A, B, C, D and A', B', C', D' are the 4 vertices of squares of length 2, respectively. Now, the circumscribed circles

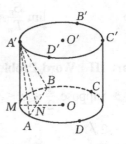

with centers O and O' of the squares constitute the bases of another cylinder, and the projecting point of A' on the bottom base is the middle point M of arc AB.

In $\triangle A'AB$, we have $A'A = A'B = AB = 2$, then $A'N = \sqrt{3}$, where N is the middle point of AB. Meanwhile, $OM = OA = \sqrt{2}$, $ON = 1$, so

$$MN = \sqrt{2} - 1, \quad A'M = \sqrt{(A'N)^2 - (MN)^2} = \sqrt[4]{8}.$$

Then the height of the original cylinder is $\sqrt[4]{8} + 2$.

Remark In order to solve the problem, you must first be clear about the way the balls are placed (each ball in contact with 4 other balls), then determine the positions of the centers of the balls.

🔟 Let $M_n = \{0. a_1 a_2 \cdots a_n \mid a_i = 0 \text{ or } 1, 1 \leqslant i \leqslant n-1, a_n = 1\}$ be a set of decimal fractions, T_n and S_n be the number and the sum of the elements in M_n respectively. Then

$$\lim_{n \to \infty} \frac{S_n}{T_n} = \underline{\hspace{2cm}}.$$

Solution Since a_1, a_2, $\cdots$, a_{n-1} all have exactly two possible values, so $T_n = 2^{n-1}$. Meanwhile, the frequency of $a_i = 1$ is the same as that of $a_i = 0$ for $1 \leqslant i \leqslant n-1$, and $a_n = 1$. Then

$$S_n = \frac{1}{2} \times 2^{n-1} \times \left(\frac{1}{10} + \frac{1}{10^2} + \cdots + \frac{1}{10^{n-1}} \right) + 2^{n-1} \times \frac{1}{10^n}$$

$$= 2^{n-1} \times \frac{1}{18} \times \left(1 - \frac{1}{10^{n-1}} \right) + 2^{n-1} \times \frac{1}{10^n}.$$

So
$$\lim_{n\to\infty}\frac{S_n}{T_n}=\lim_{n\to\infty}\left[\frac{1}{18}\left(1-\frac{1}{10^{n-1}}\right)+\frac{1}{10^n}\right]=\frac{1}{18}.$$

Part III   Word Problems (Questions 13 to 15 carry 20 marks each.)

⑬ Suppose $\frac{3}{2}\leqslant x\leqslant 5$. Prove that $2\sqrt{x+1}+\sqrt{2x-3}+\sqrt{15-3x}<2\sqrt{19}$.

Solution   By Cauchy's inequality, we have

$$2\sqrt{x+1}+\sqrt{2x-3}+\sqrt{15-3x}$$
$$=\sqrt{x+1}+\sqrt{x+1}+\sqrt{2x-3}+\sqrt{15-3x}$$
$$\leqslant\sqrt{[(x+1)+(x+1)+(2x-3)+(15-3x)](1^2+1^2+1^2+1^2)}$$
$$=2\sqrt{x+14}\leqslant 2\sqrt{19},$$

and the equality holds if and only $\sqrt{x+1}=\sqrt{2x-3}=\sqrt{15-3x}$ and $x=5$. But this is impossible. So $2\sqrt{x+1}+\sqrt{2x-3}+\sqrt{15-3x}<2\sqrt{19}$.

Remark   Some student contestants used the estimate that

$$2\sqrt{x+1}+\sqrt{2x-3}+\sqrt{15-3x}$$
$$\leqslant\sqrt{[(x+1)+(2x-3)+(15-3x)](2^2+1^2+1^2)}$$
$$=\sqrt{78},$$

but this does not give the value as required.

⑭ Suppose A, B, C are three non-collinear points corresponding to complex numbers $z_0=ai$, $z_1=\frac{1}{2}+bi$, $z_2=1+ci$ (a, b and c being real numbers), respectively. Prove that the curve

$$z=z_0\cos^4 t+2z_1\cos^2 t\cdot\sin^2 t+z_2\sin^4 t\ (t\in\mathbf{R})$$

shares a single common point with the line bisecting AB and parallel to AC in $\triangle ABC$, and find this point.

Solution Ⅰ   Let $z = x + y\mathrm{i}$ (x, $y \in \mathbf{R}$), then

$$x + y\mathrm{i} = a\cos^4 t \cdot \mathrm{i} + 2\left(\frac{1}{2} + b\mathrm{i}\right)\cos^2 t \cdot \sin^2 t + (1 + c\mathrm{i})\sin^4 t.$$

Separating real and imaginary parts, we get

$$x = \cos^2 t \cdot \sin^2 t + \sin^4 t = \sin^2 t,$$

$$y = a(1-x)^2 + 2b(1-x)x + cx^2$$

$$(0 \leqslant x \leqslant 1).$$

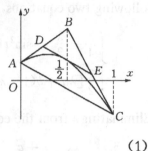

That is, $y = (a + c - 2b)x^2 + 2(b - a)x +$

$$a \quad (0 \leqslant x \leqslant 1) \tag{1}$$

Since A, B, C are non-collinear, $a + c - 2b \neq 0$. So Equation (1) is the segment of a parabola (see the diagram). Furthermore, the midpoints of AB and BC are $D\left(\frac{1}{4}, \frac{a+b}{2}\right)$ and $E\left(\frac{3}{4}, \frac{b+c}{2}\right)$, respectively. So

the equation of line DE is $y = (c - a)x + \frac{1}{4}(3a + 2b - c)$.   (2)

Solving Equations (1) and (2) simultaneously, we get $(a + c - 2b)\left(x - \frac{1}{2}\right)^2 = 0$. Then $x = \frac{1}{2}$, since $a + c - 2b \neq 0$. So the parabola and

line DE have one and only one common point $P\left(\frac{1}{2}, \frac{a+c+2b}{4}\right)$. Notice

that $\frac{1}{4} < \frac{1}{2} < \frac{3}{4}$, so point P is on the segment DE and satisfies

Equation (1), as required.

Solution Ⅱ   We can solve the problem using the method of complex numbers directly. Let D, E be the midpoints of AB, CB, respectively.

Then the complex numbers corresponding to D, E are $\frac{1}{2}(z_0 + z_1) =$

$\frac{1}{4} + \frac{a+b}{2}\mathrm{i}$, $\frac{1}{2}(z_1 + z_2) = \frac{3}{4} + \frac{b+c}{2}\mathrm{i}$, respectively. So, complex

number z corresponding to a point on the segment DE satisfies

$$z = \lambda\left(\frac{1}{4} + \frac{a+b}{2}\mathrm{i}\right) + (1 - \lambda)\left(\frac{3}{4} + \frac{b+c}{2}\mathrm{i}\right), \quad 0 \leqslant \lambda \leqslant 1.$$

Substitute the above expression into the equation of the curve

$$z = z_0 \cos^4 t + 2z_1 \cos^2 t \cdot \sin^2 t + z_2 \sin^4 t,$$

and separate the real and imaginary parts from both sides to give the following two equations,

$$\begin{cases} \dfrac{3}{4} - \dfrac{\lambda}{2} = \sin^2 t \cos^2 t + \sin^4 t, \\ \dfrac{1}{2} [\lambda a + b(1-\lambda)c] = a\cos^4 t + 2b\sin^2 t \cos^2 t + c\sin^4 t. \end{cases}$$

Eliminating λ from the equations, we get

$$\frac{3}{4}(a-c) + \frac{b+c}{2} = a\cos^4 t + (2b+a-c)\sin^2 t \cos^2 t + a\sin^4 t$$

$$= a(1 - 2\sin^2 t \cos^2 t) + (2b+a-c)\sin^2 t \cos^2 t$$

$$= a + (2b-a-c)\sin^2 t \cos^2 t.$$

Then $(2b-a-c)\left(\sin^2 t \cos^2 t - \dfrac{1}{4}\right) = 0$. Since A, B, C are non-collinear, we know that $z_1 \neq \dfrac{1}{2}(z_0 + z_2)$. So $2b-a-c \neq 0$. Then $\sin^2 t \cos^2 t = \sin^2 t(1-\sin^2 t) = \dfrac{1}{4}$, that is, $\left(\sin^2 t - \dfrac{1}{2}\right)^2 = 0$. Then we have $\dfrac{3}{4} - \dfrac{\lambda}{2} = \dfrac{1}{4} + \left(\dfrac{1}{2}\right)^2 = \dfrac{1}{2}$, so $\lambda = \dfrac{1}{2} \in [0, 1]$. That means that the curve and the line DE have one and only one common point, and the complex number corresponding to this common point is

$$z = \frac{1}{2}\left(\frac{1}{4} + \frac{a+b}{2}i\right) + \frac{1}{2}\left(\frac{3}{4} + \frac{b+c}{2}i\right) = \frac{1}{2} + \frac{a+c+2b}{4}i.$$

⑮ A circle with center O and radius R is drawn on a paper, and A is a given point in the circle with $OA = a$. Fold the paper to make a point A' on the circumference coincident with point A, then a crease line is left on the paper. Find out the set of all points on such crease lines, when A' goes through every point on the circumference.

Solution　Establish an xy-coodinate system as in the diagram with $A(a, 0)$ given. Then the crease line MN is the perpendicular bisector of segment AA' when A' $(R\cos a, R\sin a)$ is made coincident with A by folding the paper. Let $P(x, y)$ be any point on MN, then $|PA'| = |PA|$. That is

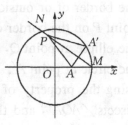

$$(x - R\cos a)^2 + (y - R\sin a)^2 = (x - a)^2 + y^2.$$

Then
$$\frac{x\cos x + y\sin a}{\sqrt{x^2 + y^2}} = \frac{R^2 - a^2 + 2ax}{2R\sqrt{x^2 + y^2}}.$$

We get
$$\sin(\theta + a) = \frac{R^2 - a^2 + 2ax}{2R\sqrt{x^2 + y^2}},$$

where
$$\sin\theta = \frac{x}{\sqrt{x^2 + y^2}}, \quad \cos\theta = \frac{y}{\sqrt{x^2 + y^2}}.$$

So
$$\left| \frac{R^2 - a^2 + 2ax}{2R\sqrt{x^2 + y^2}} \right| \leqslant 1.$$

Squaring both sides, we get $\dfrac{(2x - a)^2}{R^2} + \dfrac{4y^2}{R^2 - a^2} \geqslant 1$.

So the set we want consists of all of the points on the border of or outside the ellipse $\dfrac{(2x - a)^2}{R^2} + \dfrac{4y^2}{R^2 - a^2} = 1$.

Remark　As seen in the diagram, suppose the crease line intersects OA' at point Q. Then from $QA = QA'$ we have $OQ + QA = OA' = R$; that is, Q is on an ellipse whose foci are O and A. And the expression of the ellipse is

$$\frac{\left(x - \dfrac{a}{2}\right)^2}{\left(\dfrac{R}{2}\right)^2} + \frac{y^2}{\left(\dfrac{1}{2}\sqrt{R^2 - a^2}\right)^2} = 1.$$

For any other point P on the crease line MN, we always have $PO + PA = PO + PA' > OA' = R$. So a point of the crease line is either on the border of or outside the ellipse. On the other hand, if from any point P on the border or outside the ellipse, we draw a line tangent to the ellipse at point Q, then $QO + QA = R$. Suppose line QO intersects the circle at point A'. Then $QO + QA' = R$, that is, $QA = QA'$. Then using the property of the tangent to an ellipse, we have that PQ bisects $\angle AQA'$, and that means the tangent PQ is the crease line as mentioned above. The arguments above reveal that the set we find consists of all of the points on every tangent to the ellipse.

2004 (Hainan)

Popularization Committee of CMS and Hainan Mathematical Society were responsible for the assignment of the competition problems in the first and the second rounds of the contests.

Part I   Multiple-choice Questions (Questions 1 to 6 carry 6 marks each.)

① Let θ be an acute angle such that the equation $x^2 + 4x\cos\theta + \cot\theta = 0$ involving variable x has multiple roots. Then the measure of θ in radians is (   ).

(A) $\dfrac{\pi}{6}$ (B) $\dfrac{\pi}{12}$ or $\dfrac{5\pi}{12}$ (C) $\dfrac{\pi}{6}$ or $\dfrac{5\pi}{12}$ (D) $\dfrac{\pi}{12}$

Solution   Since the equation $x^2 + 4x\cos\theta + \cot\theta = 0$ has multiple roots, we have

$$\Delta = 16\cos^2\theta - 4\cot\theta = 0,$$

or

$$4\cot\theta(2\sin 2\theta - 1) = 0.$$

By assumption, $0 < \theta < \dfrac{\pi}{2}$, we get

$$\sin 2\theta = \frac{1}{2}.$$

It follows that

$$2\theta = \frac{\pi}{6} \text{ or } 2\theta = \frac{5\pi}{6}.$$

Thus $\theta = \frac{\pi}{12}$ or $\theta = \frac{5\pi}{12}$. Answer: B.

② Assume that $M = \{(x, y) \mid x^2 + 2y^2 = 3\}$, and $N = \{(x, y) \mid y = mx + b\}$. If $M \cap N \neq \varnothing$ for all $m \in \mathbf{R}$, then b takes values from ( ).

(A) $\left[-\frac{\sqrt{6}}{2}, \frac{\sqrt{6}}{2}\right]$ (B) $\left(-\frac{\sqrt{6}}{2}, \frac{\sqrt{6}}{2}\right)$

(C) $\left(-\frac{\sqrt{6}}{2}, \frac{\sqrt{6}}{2}\right]$ (D) $\left[-\frac{2\sqrt{3}}{3}, \frac{2\sqrt{3}}{3}\right]$

Solution For any $m \in \mathbf{R}$ we have $M \cap N \neq \varnothing$, which means point $(0, b)$ is on or in the ellipsoid $\frac{x^2}{3} + \frac{2y^2}{3} = 1$. Therefore

$$\frac{2b^2}{3} \leqslant 1, \text{ or } -\frac{\sqrt{6}}{2} \leqslant b \leqslant \frac{\sqrt{6}}{2}.$$

Answer: A.

③ The solution set of the inequality $\sqrt{\log_2 x - 1} + \frac{1}{2}\log_{\frac{1}{2}} x^3 + 2 > 0$ is ( ).

(A) $[2, 3)$ (B) $(2, 3]$ (C) $[2, 4)$ (D) $(2, 4]$

Solution The initial inequality is equivalent to

$$\begin{cases} \sqrt{\log_2 x - 1} - \frac{3}{2}\log_2 x + \frac{3}{2} + \frac{1}{2} > 0, \\ \log_2 x - 1 \geqslant 0. \end{cases}$$

Let $t = \sqrt{\log_2 x - 1}$, we have

$$\begin{cases} t - \frac{3}{2}t^2 + \frac{1}{2} > 0, \\ t \geqslant 0. \end{cases}$$

The solution of the above inequalities is $0 \leqslant t < 1$, or $0 \leqslant \log_2 x - 1 < 1$, which implies that $2 \leqslant x < 4$. Answer: C.

4 Let O be an interior point of $\triangle ABC$ such that $\overrightarrow{OA} + 2\overrightarrow{OB} + 3\overrightarrow{OC} = 0$. Then the ratio of the area of $\triangle ABC$ to the area of $\triangle AOC$ is ( ).

(A) 2   (B) $\dfrac{3}{2}$ (C) 3   (D) $\dfrac{5}{3}$

Solution   In the diagram, let D and E be the midpoints of the sides AC and BC, respectively. Then we have

$$\overrightarrow{OA} + \overrightarrow{OC} = 2\overrightarrow{OD}, \qquad (1)$$

and

$$2(\overrightarrow{OB} + \overrightarrow{OC}) = 4\overrightarrow{OE}. \qquad (2)$$

By equations (1) and (2) we get

$$\overrightarrow{OA} + 2\overrightarrow{OB} + 3\overrightarrow{OC} = 2(\overrightarrow{OD} + 2\overrightarrow{OE}) = 0.$$

It follows that $\overrightarrow{OD}$ and $\overrightarrow{OE}$ are collinear, and

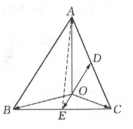

$|\overrightarrow{OD}| = 2|\overrightarrow{OE}|$. Consequently, $\dfrac{S_{\triangle AEC}}{S_{\triangle AOC}} = \dfrac{3}{2}$

and $\dfrac{S_{\triangle ABC}}{S_{\triangle AOC}} = \dfrac{3 \times 2}{2} = 3$. Answer: C.

5 Let $n = \overline{abc}$ be a 3-digit number. If we can construct an isosceles triangle (including equilateral triangle) with a, b and c as the lengths of the sides. The number of such 3-digit integers n is ( ).

(A) 45   (B) 81   (C) 165   (D) 216

Solution   If a, b and c are the lengths of the sides of a triangle, all of them are not zero, it follows that a, b, $c \in \{1, 2, \cdots, 9\}$.

(i) If the triangle we construct is equilateral, let n_1 be the number of such 3-digit numbers. Since the three digits in such 3-digit number are equal, we have

$$n_1 = C_9^1 = 9.$$

(ii) If the triangle we construct is isosceles but not equilateral, let n_2 be the number of such 3-digit numbers. Since there are only 2 different digits in such a 3-digit number, denote them by a and b. Note that the equal sides and the base of an isosceles triangle can be replaced by each other, thus the number of such pairs (a, b) is $2C_9^2$. But if the bigger number, say a, is the length of the base, then a must satisfy the condition $b < a < 2b$. All pairs that do not satisfy this condition we list in the following table. There are 20 pairs.

a	9	8	7	6	5	4	3	2	1
b	4, 3, 2, 1	4, 3, 2, 1	3, 2, 1	3, 2, 1	2, 1	2, 1	1	1	

On the other hand, there are C_3^2 possible 3-digit numbers with digits taken from a given pair (a, b). Thus

$$n_2 = C_3^2(2C_9^2 - 20) = 6(C_9^2 - 10) = 156.$$

Consequently, $n = n_1 + n_2 = 165$. Answer: C.

6 The vertical cross-section of a circular cone with vertex P is an isosceles right-angled triangle. Point A is on the circumference of the base circle, point B is interior to the base circle, O is the center of the base circle, $AB \perp OB$ and intersecting at B, $OH \perp PB$ and intersecting at H, $PA = 4$, and C is the midpoint of PA. When the tetrahedron $O - HPC$ has the maximum volume, the length of OB is (   ).

    (A) $\dfrac{\sqrt{5}}{3}$ (B) $\dfrac{2\sqrt{5}}{3}$ (C) $\dfrac{\sqrt{6}}{3}$ (D) $\dfrac{2\sqrt{6}}{3}$

Solution   Since $AB \perp OB$, and $AB \perp OP$, we have $AB \perp PB$, and $PAB \perp POB$. Moreover, from $OH \perp PB$ we obtain that $OH \perp HC$ and $OH \perp PA$. Since C is the midpoint of PA, $OC \perp PA$. Thus, PC is the altitude of the

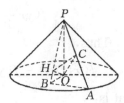

tetrahedron $O-HPC$ and $PC = 2$.

In Rt$\triangle$OHC, $OC = 2$. Therefore when $HO = HC$, $S_{\triangle ABC}$ reaches its maximum, that is, $V_{O-HPC} = V_{P-HCO}$ reaches its maximum. In this case, $HO = \sqrt{2}$, and $HO = \dfrac{1}{2}OP$. Hence, $\angle HPO = 30°$, and $OB = OP \cdot \tan 30° = \dfrac{2\sqrt{6}}{3}$. Answer: D.

Part II   Short-answer Questions (Questions 7 to 12 carry 6 marks each.)

⑦ In a planar rectangular coordinate system xOy, the area enclosed by the graph of function $f(x) = a\sin ax + \cos ax$ $(a > 0)$ defined on an interval with the least positive period and by the graph of function $g(x) = \sqrt{a^2 + 1}$ is _____.

Solution   We rewrite function $f(x)$ as $f(x) = \sqrt{a^2 + 1}\sin(ax + \varphi)$, where $\varphi = \arctan\dfrac{1}{a}$. Its least positive period is $\dfrac{2\pi}{a}$, and its amplitude is $\sqrt{a^2 + 1}$. By symmetry of the figure enclosed by the graphs of the functions $f(x)$ and $g(x)$, we can change the figure into a rectangle with length $\dfrac{2\pi}{a}$ and width $\sqrt{a^2 + 1}$ using the cut-and-paste method. Therefore its area is $\dfrac{2\pi}{a}\sqrt{a^2 + 1}$.

⑧ Let $f: \mathbf{R} \rightarrow \mathbf{R}$ be a function such that $f(0) = 1$ and for any x, $y \in \mathbf{R}$, $f(xy+1) = f(x)f(y) - f(y) - x + 2$ holds. Then $f(x) = $ _____.

Solution   Since for any x, $y \in \mathbf{R}$, $f(xy+1) = f(x)f(y) - f(y) - x + 2$, we have

$$f(yx + 1) = f(y)f(x) - f(x) - y + 2.$$

Thus,

$$f(x)f(y) - f(y) - x + 2 = f(y)f(x) - f(x) - y + 2,$$

that is,

$$f(x) + y = f(y) + x.$$

Put $y = 0$, we obtain $f(x) = x + 1$.

⑨ In the diagram, $ABCD - A_1B_1C_1D_1$ is a cube. The dihedral angle $A - BD_1 - A_1$ in degrees is _____.

Solution   Draw line segments D_1C, and CE such that $CE \perp BD_1$, E is the foot of the perpendicular. Let the extended lines CE and A_1B intersect at F. Draw line segment AE. By symmetry we have $AE \perp BD_1$. Therefore $\angle FEA$ is the plane angle of the dihedral angle $A - BD_1 - A_1$. Draw a line segment connecting points A and C, and set

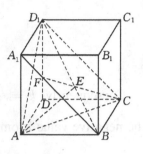

$AB = 1$, then $AC = AD_1 = \sqrt{2}$, $BD_1 = \sqrt{3}$. In $\mathrm{Rt}\triangle ABD_1$, $AE = \dfrac{AB \cdot AD_1}{BD_1} = \dfrac{\sqrt{2}}{\sqrt{3}}$. In $\triangle AEC$,

$$\cos\angle AEC = \frac{AE^2 + CE^2 - AC^2}{2AE \cdot CE} = \frac{2AE^2 - AC^2}{2AE^2}$$

$$= \frac{\dfrac{4}{3} - 2}{\dfrac{4}{3}} = -\frac{1}{2}.$$

Thus $\angle AEC = 120°$. But $\angle FEA$ is the supplementary angle of $\angle AEC$, hence $\angle FEA = 60°$.

⑩ Let p be an odd prime. Let k be a positive integer such that $\sqrt{k^2 - pk}$ is also a positive integer. Then $k = $ _____.

Solution   Set $\sqrt{k^2 - pk} = n$, $n \in \mathbf{N}$. Thus $k^2 - pk - n^2 = 0$, and $k = \dfrac{p \pm \sqrt{p^2 + 4n^2}}{2}$, which implies that $p^2 + 4n^2$ is a perfect square, says m^2, where $m \in \mathbf{N}$. So $(m - 2n)(m + 2n) = p^2$.

Since p is a prime and $p \geqslant 3$, we have

$$\begin{cases} m - 2n = 1, \\ m + 2n = p^2. \end{cases}$$

Solve the equations above and we get

$$\begin{cases} m = \dfrac{p^2 + 1}{2}, \\ n = \dfrac{p^2 - 1}{4}. \end{cases}$$

Consequently, $k = \dfrac{p \pm m}{2} = \dfrac{2p \pm (p^2 + 1)}{4}$. Thus $k = \dfrac{(p+1)^2}{4}$
(the negative value is omitted).

Let a_0, a_1, a_2, $\cdots$, a_n, $\cdots$ be a sequence of numbers satisfying $(3 - a_{n+1}) \cdot (6 + a_n) = 18$, and $a_0 = 3$. Then $\displaystyle\sum_{i=0}^{n} \frac{1}{a_i}$ equals _____.

Solution   Set $b_n = \dfrac{1}{a_n}$, $n = 0, 1, 2, \cdots$, then $\left(3 - \dfrac{1}{b_{n+1}}\right)\left(6 + \dfrac{1}{b_n}\right) = 18$, namely,

$$3b_{n+1} - 6b_n - 1 = 0.$$

Hence $b_{n+1} = 2b_n + \dfrac{1}{3}$, or $b_{n+1} + \dfrac{1}{3} = 2\left(b_n + \dfrac{1}{3}\right)$. So $\left\{b_n + \dfrac{1}{3}\right\}$ is a geometric progression with common ratio 2. Thus

$$b_n + \frac{1}{3} = 2^n\left(b_0 + \frac{1}{3}\right) = 2^n\left(\frac{1}{a_0} + \frac{1}{3}\right) = \frac{1}{3} \times 2^{n+1},$$

$$b_n = \frac{1}{3}(2^{n+1} - 1).$$

Therefore,

$$\sum_{i=0}^{n} \frac{1}{a_i} = \sum_{i=0}^{n} b_i = \sum_{i=0}^{n} \frac{1}{3}(2^{i+1} - 1)$$

$$= \frac{1}{3}\left[\frac{2(2^{n+1} - 1)}{2 - 1} - (n+1)\right]$$

$$= \frac{1}{3}(2^{n+2} - n - 3).$$

⑫ Let $M(-1, 2)$ and $N(1, 4)$ be two points in a plane rectangular coordinate system xOy. P is a moving point on the x-axis. When $\angle MPN$ takes its maximum value, the x-coordinate of point P is _____.

Solution The center of a circle passing through points M and N is on the perpendicular bisector $y = 3 - x$ of MN. Denote the center by $S(a, 3-a)$, then the equation of the circle S is

$$(x - a)^2 + (y - 3 + a)^2 = 2(1 + a^2).$$

Since for a chord with a fixed length, the angle at the circumference subtended by the corresponding arc will become larger as the radius of the circle becomes smaller. When $\angle MPN$ reaches its maximum value, the circle S through the three points M, N and P will be tangent to the x-axis at P, which means the value a in the equation of S has to satisfy the condition $2(1 + a^2) = (a - 3)^2$. Solve the above equation we have $a = 1$ or $a = -7$. Thus the points of contact are $P(1, 0)$ and $P'(-7, 0)$ respectively.

But the radius of the circle through points M, N, and P' is larger than that of the circle through points M, N and P. Therefore $\angle MPN > \angle MP'N$. Thus $P(1, 0)$ is the point we want to find, and the x-axis of point P is 1.

Part III Word Problems (Questions 13 to 15 carry 20 marks each.)

⑬ The rule of an "obstacle course" specifies that at the nth obstacle a person has to toss a die n times. If the sum of points in these n tosses is bigger than 2^n, the person is said to have crossed the obstacle.

   (1) At most how many obstacles can a person cross?
   (2) What is the probability that a person crosses the first three obstacles?

(Note: A die is a fair regular cube, on its six faces there are numbers 1, 2, 3, 4, 5, 6 respectively. Toss a die, the point is the number appearing on its top face after it stops moving.)

Solution Since the die is fair, the probability of any of the six numbers appearing is the same.

(1) Since the highest point of a die is 6, and $6 \times 4 > 2^4$, $6 \times 5 < 2^5$, it is impossible that the sum of points appearing in n tosses is bigger than 2^n if $n \geqslant 5$. This means it is an impossible event, and the probability of crossing the obstacle is 0.

Therefore at most 4 obstacles that a person can cross.

(2) We denote A_n the event "at the nth obstacle the person fails to cross", the complementary event $\overline{A_n}$ is "at the nth obstacle the person crosses successfully".

At the nth obstacle of this game the number of all possible outcomes is 6^n.

The first obstacle: event A_1 contains 2 possible outcomes (i. e., the outcomes in which the number appearing is 1 or 2). So the probability of crossing the obstacle is

$$P(\overline{A_1}) = 1 - P(A_1) = 1 - \frac{2}{6} = \frac{2}{3}.$$

The second obstacle: the number of outcomes contained in event A_2 is the total number of positive integer solution sets of the equation $x + y = a$ where a is taken to be 2, 3 and 4 respectively. Thus the number of outcomes equals $C_1^1 + C_2^1 + C_3^1 = 1 + 2 + 3 = 6$, and the probability of crossing the obstacle is

$$P(\overline{A_2}) = 1 - P(A_2) = 1 - \frac{6}{6^2} = \frac{5}{6}.$$

The third obstacle: the number of outcomes contained in event A_3 is the total number of positive integer solution sets of the equation $x + y + z = a$ where a is taken to be 3, 4, 5, 6, 7 and 8 respectively. Thus the number of outcomes equals

$$C_2^2 + C_3^2 + C_4^2 + C_5^2 + C_6^2 + C_7^2 = 1 + 3 + 6 + 10 + 15 + 21 = 56,$$

and the probability of crossing the obstacle is

$$P(\overline{A_3}) = 1 - P(A_3) = 1 - \frac{56}{6^3} = \frac{20}{27}.$$

Consequently, the probability that a person crosses the first three obstacles is

$$P(\overline{A_1}) \times P(\overline{A_2}) \times P(\overline{A_3}) = \frac{2}{3} \times \frac{5}{6} \times \frac{20}{27} = \frac{100}{243}.$$

(We can also list all the possible outcomes at the second obstacle and at the third obstacle.)

Remark   Problems concerning probability theory first appeared in the National High School Mathematics Competition. Problems are not too difficult. Topics such as derivative and its applications have already appeared in high school textbooks. These topics will also appear in mathematics competitions.

⑫ In a plane rectangular coordinate system xOy there are three points $A\left(0, \frac{4}{3}\right)$, $B(-1, 0)$ and $C(1, 0)$. The distance from point P to line BC is the geometric mean of the distances from this point to lines AB and AC.

(1) Find the locus equation of point P.

(2) If line L passes through the incenter (say, D) of $\triangle ABC$, and has exactly 3 common points with the locus of point P. Determine all values of the slope k of line L.

Solution   (1) The equations of lines AB, AC and BC are $y = \frac{4}{3}(x + 1)$, $y = -\frac{4}{3}(x - 1)$ and $y = 0$ respectively. The distances from point P to AB, AC and BC are respectively

$$d_1 = \frac{1}{5} \mid 4x - 3y + 4 \mid,$$

$$d_2 = \frac{1}{5} \mid 4x + 3y - 4 \mid, \ d_3 = \mid y \mid.$$

According to the assumption, $d_1 d_2 = d_3^2$, we have $\mid 16x^2 - (3y-4)^2 \mid = 25y^2$. That is

$$16x^2 - (3y-4)^2 + 25y^2 = 0, \text{ or } 16x^2 - (3y-4)^2 - 25y^2 = 0.$$

By simplifying the above equations, we obtain that the locus equations of point P consist of

$$\text{circle } S\colon 2x^2 + 2y^2 + 3y - 2 = 0$$

and

$$\text{hyperbola } T\colon 8x^2 - 17y^2 + 12y - 8 = 0.$$

(2) According to (1), the locus of point P consists of two parts

$$\text{circle } S\colon 2x^2 + 2y^2 + 3y - 2 = 0, \qquad \qquad ①$$

and

$$\text{hyperbola } T\colon 8x^2 - 17y^2 + 12y - 8 = 0. \qquad \qquad ②$$

Since $B(-1, 0)$ and $C(1, 0)$ are points satisfying the assumption, points B and C are on the locus of point P, and the common points of the curves S and T are points B and C only.

The incenter of $\triangle ABC$ is also a point satisfying the assumption. In view of $d_1 = d_2 = d_3$, solving the equations we have $D\left(0, \frac{1}{2}\right)$.

Line L passes though D, and has three common points with the locus of point P. So the slope of L is defined. Suppose that the equation of L is

$$y = kx + \frac{1}{2}. \qquad \qquad ③$$

(i) If $k = 0$, then L is tangent to the circle S, which means there is a unique common point D. In this case line L is parallel to the x-axis, which implies that L and hyperbola T have two other common points different from point D. Hence, there are just three common points for L and the locus of point P.

(ⅱ) If $k \neq 0$, there are two different intersection points for L and the circle S. In order that L and the locus of point P have 3 common points, we must have one of the following two cases.

Case 1: Line L passes through point B or point C, which means that the slope of L is $k = \pm \dfrac{1}{2}$, and the equation of L is $x = \pm (2y-1)$. Substitute it into equation ② we get

$$y(3y - 4) = 0.$$

Solving it we have $E\left(\dfrac{5}{3}, \dfrac{4}{3}\right)$ or $F\left(-\dfrac{5}{3}, \dfrac{4}{3}\right)$, which means that line BD and curve T have 2 intersection points B and E, and line CD and curve T have 2 intersection points C and F.

Consequently, if $k = \pm \dfrac{1}{2}$ for L and the locus of point P there are exactly 3 common points.

Case 2: Line L does not pass through point B and point C $\left(\text{i. e. } k \neq \pm \dfrac{1}{2}\right)$. Since for L and S there are two different intersection points, there exists a unique common point for L and hyperbola T. Thus for the following system of equations

$$\begin{cases} 8x^2 - 17y^2 + 12y - 8 = 0, \\ y = kx + \dfrac{1}{2}, \end{cases}$$

there is one and only one real solution. After eliminating y and simplifying we have

$$(8 - 17k^2)x^2 - 5kx - \dfrac{25}{4} = 0.$$

The above equation has a unique real solution if and only if

$$8 - 17k^2 = 0 \qquad\qquad ④$$

or

$$(-5k)^2 + 4(8 - 17k^2) \frac{25}{4} = 0. \qquad\qquad ⑤$$

Solving equation ④ we get $k = \pm \frac{2\sqrt{34}}{17}$. And solving equation ⑤

we obtain $k = \pm \frac{\sqrt{2}}{2}$.

Consequently, the set of all possible values of the slope k of line L is the following finite set

$$\left\{ 0, \pm \frac{1}{2}, \pm \frac{2\sqrt{34}}{17}, \pm \frac{\sqrt{2}}{2} \right\}.$$

⑮ Suppose that α and β are different real roots of the equation $4x^2 - 4tx - 1 = 0$ ($t \in \mathbf{R}$). $[\alpha, \beta]$ is the domain of the function $f(x) = \dfrac{2x - t}{x^2 + 1}$.

(1) Find $g(t) = \max f(x) - \min f(x)$.

(2) Prove that for $u_i \in \left(0, \dfrac{\pi}{2}\right)$ ($i = 1, 2, 3$), if $\sin u_1 + \sin u_2 + \sin u_3 = 1$, then

$$\frac{1}{g(\tan u_1)} + \frac{1}{g(\tan u_2)} + \frac{1}{g(\tan u_3)} < \frac{3}{4}\sqrt{6}.$$

Solution (1) Let $\alpha \leqslant x_1 < x_2 \leqslant \beta$, then

$$4x_1^2 - 4tx_1 - 1 \leqslant 0, \quad 4x_2^2 - 4tx_2 - 1 \leqslant 0.$$

Therefore,

$$4(x_1^2 + x_2^2) - 4t(x_1 + x_2) - 2 \leqslant 0,$$

$$2x_1 x_2 - t(x_1 + x_2) - \frac{1}{2} < 0.$$

But

$$f(x_2) - f(x_1) = \frac{2x_2 - t}{x_2^2 + 1} - \frac{2x_1 - t}{x_1^2 + 1}$$

$$= \frac{(x_2 - x_1)[t(x_1 + x_2) - 2x_1 x_2 + 2]}{(x_2^2 + 1)(x_1^2 + 1)},$$

and $t(x_1 + x_2) - 2x_1 x_2 + 2 > t(x_1 + x_2) - 2x_1 x_2 + \frac{1}{2} > 0$, thus

$$f(x_2) - f(x_1) > 0.$$

Consequently, $f(x)$ is an increasing function on the interval $[\alpha, \beta]$.

Since $\alpha + \beta = t$ and $\alpha\beta = -\frac{1}{4}$,

$$g(t) = \max f(x) - \min f(x) = f(\beta) - f(\alpha)$$

$$= \frac{\sqrt{t^2 + 1}\left(t^2 + \frac{5}{2}\right)}{t^2 + \frac{25}{16}} = \frac{8\sqrt{t^2 + 1}(2t^2 + 5)}{16t^2 + 25}.$$

(2) $g(\tan u_i) = \dfrac{\dfrac{8}{\cos u_i}\left(\dfrac{2}{\cos^2 u_i} + 3\right)}{\dfrac{16}{\cos^2 u_i} + 9} = \dfrac{\dfrac{16}{\cos u_i} + 24\cos u_i}{16 + 9\cos^2 u_i}$

$$\geqslant \frac{2\sqrt{16 \times 24}}{16 + 9\cos^2 u_i} = \frac{16\sqrt{6}}{16 + 9\cos^2 u_i} (i = 1, 2, 3),$$

so

$$\sum_{i=1}^{3} \frac{1}{g(\tan u_i)} \leqslant \frac{1}{16\sqrt{6}} \sum_{i=1}^{3} (16 + 9\cos^2 u_i)$$

$$= \frac{1}{16\sqrt{6}}\left(16 \times 3 + 9 \times 3 - 9\sum_{i=1}^{3} \sin^2 u_i\right).$$

Since $\sum_{i=1}^{3} \sin u_i = 1$, and $u_i \in \left(0, \frac{\pi}{2}\right)$, $i = 1, 2, 3$, we obtain

$$3\sum_{i=1}^{3} \sin^2 u_i > \left(\sum_{i=1}^{3} \sin u_i\right)^2 = 1.$$

Thus

$$\frac{1}{g(\tan u_1)} + \frac{1}{g(\tan u_2)} + \frac{1}{g(\tan u_3)}$$

$$< \frac{1}{16\sqrt{6}}\left(75 - 9 \times \frac{1}{3}\right) = \frac{3}{4}\sqrt{6}.$$

Remark   Part (1) of this problem is well-known, we put in an inequality to increase the level of difficulty.

2005   (Jiangxi)

Popularization Committee of CMS and Jiangxi Mathematical Society were responsible for the assignment of the competition problems in the first and the second rounds of the contests.

Part I   Multiple-choice Questions (Questions 1 to 6 carry 6 marks each.)

1 Let k be a real number such that the inequality $\sqrt{x-3} + \sqrt{6-x} \geqslant k$ has a solution. The maximum value of k is (   ).

   (A) $\sqrt{6} - \sqrt{3}$ (B) $\sqrt{3}$ (C) $\sqrt{6} + \sqrt{3}$ (D) $\sqrt{6}$

Solution   Set $y = \sqrt{x-3} + \sqrt{6-x}, 3 \leqslant x \leqslant 6$.
   Then

$$y^2 = (x-3) + (6-x) + 2\sqrt{(x-3)(6-x)}$$

$$\leqslant 2[(x-3) + (6-x)] = 6.$$

So $0 < y \leqslant \sqrt{6}$, and the maximum value of k is $\sqrt{6}$. Answer: D.

2 A, B, C, D are four points in the space and satisfy $|\overrightarrow{AB}| = 3$, $|\overrightarrow{BC}| = 7$, $|\overrightarrow{CD}| = 11$ and $|\overrightarrow{DA}| = 9$. Then $|\overrightarrow{AC}| \cdot |\overrightarrow{BD}|$ has (   ) values.

(A) only 1   (B) two

(C) four   (D) infinitely many

Solution   Note that $\overrightarrow{AB}^2 + \overrightarrow{CD}^2 = 3^2 + 11^2 = 130 = 7^2 + 9^2 = \overrightarrow{BC}^2 + \overrightarrow{DA}^2$. Since

$$\overrightarrow{AB} + \overrightarrow{BC} + \overrightarrow{CD} + \overrightarrow{DA} = \vec{0},$$

$$DA^2 = \overrightarrow{DA}^2 = (\overrightarrow{AB} + \overrightarrow{BC} + \overrightarrow{CD})^2$$

$$= AB^2 + BC^2 + CD^2 + 2(\overrightarrow{AB} \cdot \overrightarrow{BC} + \overrightarrow{BC} \cdot \overrightarrow{CD} + \overrightarrow{CD} \cdot \overrightarrow{AB})$$

$$= AB^2 - BC^2 + CD^2 + 2(\overrightarrow{BC}^2 + \overrightarrow{AB} \cdot \overrightarrow{BC} + \overrightarrow{BC} \cdot \overrightarrow{CD} + \overrightarrow{CD} \cdot \overrightarrow{AB})$$

$$= AB^2 - BC^2 + CD^2 + 2(\overrightarrow{AB} + \overrightarrow{BC}) \cdot (\overrightarrow{BC} + \overrightarrow{CD}),$$

i.e. $2|\overrightarrow{AC}| \cdot |\overrightarrow{BD}| = 2|\overrightarrow{AB} + \overrightarrow{BC}| \cdot |\overrightarrow{BC} + \overrightarrow{CD}|$

$$= AD^2 + BC^2 - AB^2 - CD^2 = 0.$$

Thus $|\overrightarrow{AC}| \cdot |\overrightarrow{BD}|$ has only one value 0. Answer: A.

④ $\triangle ABC$ is inscribed in a unit circle. The three bisectors of the angles A, B and C are extended to intersect the circle at A_1, B_1 and C_1 respectively.

Then the value of $\dfrac{AA_1 \cos \dfrac{A}{2} + BB_1 \cos \dfrac{B}{2} + CC_1 \cos \dfrac{C}{2}}{\sin A + \sin B + \sin C}$ is (   ).

(A) 2   (B) 4   (C) 6   (D) 8

Solution   Join BA_1 as shown in the diagram. Then

$$AA_1 = 2\sin\left(B + \frac{A}{2}\right) = 2\sin\left(\frac{A+B+C}{2} + \frac{B}{2} - \frac{C}{2}\right)$$

$$= 2\cos\left(\frac{B}{2} - \frac{C}{2}\right).$$

Thus $AA_1 \cos \dfrac{A}{2} = 2\cos\left(\dfrac{B}{2} - \dfrac{C}{2}\right)\cos\dfrac{A}{2}$

$$= \cos\left(\frac{\pi}{2} - C\right) + \cos\left(\frac{\pi}{2} - B\right) = \sin C + \sin B.$$

Similarly,

$$BB_1 \cos \frac{B}{2} = \sin A + \sin C, \quad CC_1 \cos \frac{C}{2} = \sin A + \sin B.$$

Therefore,

$$AA_1 \cos \frac{A}{2} + BB_1 \cos \frac{B}{2} + CC_1 \cos \frac{C}{2} = 2(\sin A + \sin B + \sin C),$$

and the original expression $= \dfrac{2(\sin A + \sin B + \sin C)}{\sin A + \sin B + \sin C} = 2.$

Answer: A.

As shown in the diagram, $ABCD\text{-}A'B'C'D'$ is a cube. Construct an arbitrary plane α perpendicular to the diagonal AC' such that α has common points with each face of the cube. Let S and L denote the area and the perimeter of the cross-section of α respectively. Then (   ).

(A) S is a fixed number and L is not

(B) S is not fixed and L is fixed

(C) Both S and L are fixed

(D) Neither S nor L is fixed

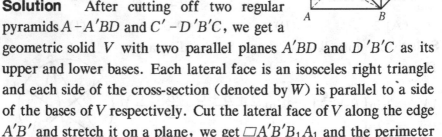

Solution   After cutting off two regular pyramids $A-A'BD$ and $C'-D'B'C$, we get a geometric solid V with two parallel planes $A'BD$ and $D'B'C$ as its upper and lower bases. Each lateral face is an isosceles right triangle and each side of the cross-section (denoted by W) is parallel to a side of the bases of V respectively. Cut the lateral face of V along the edge $A'B'$ and stretch it on a plane, we get $\square A'B'B_1A_1$ and the perimeter of W is stretched into a line segment ($E'E_1$ in the figure) which is parallel to $A'A_1$. Clearly, $E'E_1 = A'A_1$. Thus L is a fixed value.

When E' is the midpoint of $A'B'$, W is a regular 6-gon. But when E' is moved to A', W is an equilateral triangle. It is easy to see that the areas of a regular 6-gon and an equilateral triangle with the same perimeter L are $\frac{\sqrt{3}}{24}L^2$ and $\frac{\sqrt{3}}{36}L^2$ respectively. Thus S is not fixed. **Answer: B.**

⑤ The curve represented by the equation $\dfrac{x^2}{\sin\sqrt{2} - \sin\sqrt{3}} + \dfrac{y^2}{\cos\sqrt{2} - \cos\sqrt{3}} = 1$ is (   ).

  (A) An ellipse with the foci on the x-axes
  (B) A hyperbola with the foci on the x-axes
  (C) An ellipse with the foci on the y-axes
  (D) A hyperbola with the foci on the y-axes

Solution Since $\sqrt{2} + \sqrt{3} > \pi$, so $0 < \dfrac{\pi}{2} - \sqrt{2} < \sqrt{3} - \dfrac{\pi}{2} < \dfrac{\pi}{2}$ and

$$\cos\left(\frac{\pi}{2} - \sqrt{2}\right) > \cos\left(\sqrt{3} - \frac{\pi}{2}\right), \text{ i.e. } \sin\sqrt{2} > \sin\sqrt{3}.$$

Since $0 < \sqrt{2} < \dfrac{\pi}{2}$, $\dfrac{\pi}{2} < \sqrt{3} < \pi$, so $\cos\sqrt{2} > 0$, $\cos\sqrt{3} < 0$, and $\cos\sqrt{2} - \cos\sqrt{3} > 0$. Thus the curve represented by the equation is an ellipse.

Since $(\sin\sqrt{2} - \sin\sqrt{3}) - (\cos\sqrt{2} - \cos\sqrt{3})$

$$= 2\sqrt{2}\sin\frac{\sqrt{2} - \sqrt{3}}{2}\sin\left(\frac{\sqrt{2} + \sqrt{3}}{2} + \frac{\pi}{4}\right) \qquad (*)$$

and $-\dfrac{\pi}{2} < \dfrac{\sqrt{2} - \sqrt{3}}{2} < 0,$

we get $\sin\dfrac{\sqrt{2} - \sqrt{3}}{2} < 0, \ \dfrac{\pi}{2} < \dfrac{\sqrt{2} + \sqrt{3}}{2} < \dfrac{3\pi}{4},$

$$\frac{3\pi}{4} < \frac{\sqrt{2} + \sqrt{3}}{2} + \frac{\pi}{4} < \pi,$$

$$\sin\left(\frac{\sqrt{2} + \sqrt{3}}{2} + \frac{\pi}{4}\right) > 0,$$

so the expression ($*$) is less than 0.

That is $\sin\sqrt{2} - \sin\sqrt{3} < \cos\sqrt{3} - \cos\sqrt{3}$, therefore the curve is an ellipse with foci on the y-axies. Answer: C.

⑥ Let $T = \{0, 1, 2, 3, 4, 5, 6\}$ and $M = \left\{ \dfrac{a_1}{7} + \dfrac{a_2}{7^2} + \dfrac{a_3}{7^3} + \dfrac{a_4}{7^4} \right.$;

$a_i \in T$, $i = 1, 2, 3, 4 \Big\}$. Arrange the numbers in M in the descending order. Then the 2 005-th number is (   ).

(A) $\dfrac{5}{7} + \dfrac{5}{7^2} + \dfrac{6}{7^3} + \dfrac{3}{7^4}$ (B) $\dfrac{5}{7} + \dfrac{5}{7^2} + \dfrac{6}{7^3} + \dfrac{2}{7^4}$

(C) $\dfrac{1}{7} + \dfrac{1}{7^2} + \dfrac{0}{7^3} + \dfrac{4}{7^4}$ (D) $\dfrac{1}{7} + \dfrac{1}{7^2} + \dfrac{0}{7^3} + \dfrac{3}{7^4}$

Solution   Let $[a_1 a_2 \cdots a_k]_p$ be a number base p with k digits. Multiply each number in M by 7^4, and we get

$$M' = \{a_1 7^3 + a_2 7^2 + a_3 7 + a_4; \ a_i \in T,$$

$$i = 1, 2, 3, 4\} = \{[a_1 a_2 a_3 a_4]_7' \mid a_i \in T, \ i = 1, 2, 3, 4\}.$$

The maximum number in M' is $[6\,666]_7 = [2\,400]_{10}$.

In the decimal system, starting from 2 400 in the descending order, the 2 005-th number is $2\,400 - 2\,004 = 396$. But $[396]_{10} = [1\,104]_7$. Divide this number by 7^4, we get a number in M, that is $\dfrac{1}{7} + \dfrac{1}{7^2} + \dfrac{0}{7^3} + \dfrac{4}{7^4}$. Answer: C.

Part II   Short-answer Questions (Questions 7 to 12 carry 6 marks each.)

⑦ Express the polynomial in x $f(x) = 1 - x + x^2 - x^3 + \cdots - x^{19} + x^{20}$ into a polynomial in y $g(y) = a_0 + a_1 y + a_2 y^2 + \cdots + a_{19} y^{19} + a_{20} y^{20}$, where $y = x - 4$. Then $a_0 + a_1 + \cdots + a_{20} = $

_____.

Solution   The terms in the expression $f(x)$ form a geometric series with first term 1 and common ratio $-x$. By the summation formula of geometric series,

$$f(x) = \frac{(-x)^{21} - 1}{-x - 1} = \frac{x^{21} + 1}{x + 1}.$$

Set $x = y + 4$, $g(y) = \dfrac{(y+4)^{21} + 1}{y + 5}$. Let $y = 1$, we get

$$a_0 + a_1 + \cdots + a_{20} = g(1) = \frac{5^{21} + 1}{6}.$$

⑧ Let $f(x)$ be a decreasing function defined on $(0, +\infty)$. If $f(2a^2 + a + 1) < f(3a^2 - 4a + 1)$, then the range of a is ___

___ .

Solution Since $f(x)$ is defined on $(0, +\infty)$, from

$$\begin{cases} 2a^2 + a + 1 = 2\left(a + \dfrac{1}{4}\right)^2 + \dfrac{7}{8} > 0, \\ 3a^2 - 4a + 1 = (3a - 1)(a - 1) > 0, \end{cases}$$

we get $\qquad\qquad a > 1 \text{ or } a < \dfrac{1}{3}.$ (1)

Since $f(x)$ is a decreasing function on $(0, +\infty)$, so

$$2a^2 + a + 1 > 3a^2 - 4a + 1 \Rightarrow a^2 - 5a < 0.$$

Thus $0 < a < 5$. Combining this with (1), we have $0 < a < \dfrac{1}{3}$ or $1 < a < 5$.

⑨ Assume that α, β, γ satisfy $0 < \alpha < \beta < \gamma < 2\pi$. If

$$\cos(x + \alpha) + \cos(x + \beta) + \cos(x + \gamma) = 0$$

for arbitrary $x \in \mathbf{R}$, then $\gamma - \alpha = $ ___ .

Solution Write $f(x) = \cos(x + \alpha) + \cos(x + \beta) + \cos(x + \gamma)$. Since $f(x) \equiv 0$ for $x \in \mathbf{R}$,

$$f(-\alpha) = 0, \ f(-\gamma) = 0 \text{ and } f(-\beta) = 0.$$

That is
$$\cos(\beta-\alpha)+\cos(\gamma-\alpha)=-1,$$
$$\cos(\alpha-\beta)+\cos(\gamma-\beta)=-1,$$

and
$$\cos(\alpha-\gamma)+\cos(\beta-\gamma)=-1,$$

so
$$\cos(\beta-\alpha)=\cos(\gamma-\beta)=\cos(\gamma-\alpha)=-\frac{1}{2}.$$

Since $0<\alpha<\beta<\gamma<2\pi$, so $\beta-\alpha,\ \gamma-\beta,\ \gamma-\alpha\in\left\{\frac{2\pi}{3},\frac{4\pi}{3}\right\}$. In view of $\beta-\alpha<\gamma-\alpha,\ \gamma-\beta<\gamma-\alpha$, it is possible only when $\beta-\alpha=\gamma-\beta=\frac{2\pi}{3}$, so $\gamma-\alpha=\frac{4\pi}{3}$.

On the other hand, when $\beta-\alpha=\gamma-\beta=\frac{2\pi}{3}$, we have $\beta=\alpha+\frac{2\pi}{3}$, $\gamma=\alpha+\frac{4\pi}{3}$. For arbitrary $x\in\mathbf{R}$, we denote $x+\alpha=\theta$. Since three points $(\cos\theta,\ \sin\theta)$, $\left(\cos\left(\theta+\frac{2\pi}{3}\right),\ \sin\left(\theta+\frac{2\pi}{3}\right)\right)$, $\left(\cos\left(\theta+\frac{4\pi}{3}\right),\ \sin\left(\theta+\frac{4\pi}{3}\right)\right)$ are the vertices of an equilateral triangle on the unit circle $x^2+y^2=1$ with center at the origin, it is obvious that

$$\cos\theta+\cos\left(\theta+\frac{2\pi}{3}\right)+\cos\left(\theta+\frac{4\pi}{3}\right)=0,$$

and that is $\cos(x+\alpha)+\cos(x+\beta)+\cos(x+\gamma)=0$.

⑩ As shown in the diagram, the volume of tetrahedron $DABC$ is $\frac{1}{6}$. Also, $\angle ACB=45°$, and $AD+BC+\frac{AC}{\sqrt{2}}=3$. Then $CD=$

_____ .

Solution Since $\frac{1}{3}AD\cdot\left(\frac{1}{2}BC\cdot AC\cdot\right.$

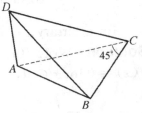

$\left.\sin 45°\right)\geqslant V_{DABC}=\frac{1}{6}$, so $AD\cdot BC\cdot\frac{AC}{\sqrt{2}}\geqslant 1.$

Note that $3 = AD + BC + \dfrac{AC}{\sqrt{2}} \geqslant 3\sqrt[3]{AD \cdot BC \cdot \dfrac{AC}{\sqrt{2}}} \geqslant 3$. The

equality holds if and only if $AD = BC = \dfrac{AC}{\sqrt{2}} = 1$. It follows that AD is

perpendicular to the face ABC. So $DC = \sqrt{AD^2 + AC^2} = \sqrt{3}$.

If one side of square $ABCD$ is on the line $y = 2x - 17$, and the other two vertices lie on parabola $y = x^2$. Then the minimum area of the square is _____ .

Solution   Assume that AB is on the line $y = 2x - 17$ and the coordinates of the other two vertices on the parabola are $C(x_1, y_1)$ and $D(x_2, y_2)$. Then CD is on a line L whose equation is $y = 2x + b$. Combining this with the equation of the parabola, we get $x^2 = 2x + b \Rightarrow x_{1, 2} = 1 \pm \sqrt{b+1}$. Assume that the length of one side of the square is a. Then

$$a^2 = (x_1 - x_2)^2 + (y_1 - y_2)^2$$

$$= 5(x_1 - x_2)^2 = 20(b+1). \tag{1}$$

Pick a point $(6, -5)$ on the line $y = 2x - 17$, and the distance from the point to the line $y = 2x + b$ is a.

So
$$a = \frac{|17 + b|}{\sqrt{5}}. \tag{2}$$

From (1) and (2), we get $b_1 = 3$, $b_2 = 63$, so $a^2 = 80$ or $a^2 = 1\,280$, and $a^2_{\min} = 80$.

A natural number a is called a "lucky number" if the sum of its digital is 7. Arrange all "lucky numbers" in an ascending order, and we get a sequence $a_1, a_2, \cdots$. If $a_n = 2\,005$, then $a_{5n} = $ _____ .

Solution   Since the number of non-negative integer solutions of equation $x_1 + x_2 + \cdots + x_k = m$ is C^m_{m+k-1}, the number of integer solutions, when $x_1 \geqslant 1$ and $x_i \geqslant 0 (i \geqslant 2)$, is C_{m+k-2}^{m-1}. Let $m = 7$, the number of lucky numbers with k digits is $p(k) = C_{k+5}^6$.

Since 2 005 is the minimum lucky number of the type $\overline{2abc}$ and $p(1) = C_6^6 = 1$, $p(2) = C_7^6 = 7$, $p(3) = C_8^6 = 28$. Note that the number of four digits lucky numbers of the type $\overline{1abc}$ is the number of non-negative integer solutions of $a+b+c = 6$, i.e. $C_{6+3-1}^6 = 28$. Thus $1 + 7 + 28 + 28 + 1 = 65$ and 2 005 is the 65-th lucky number, i.e. $a_{65} = 2\,005$, so $n = 65$, $5n = 325$.

Furthermore $p(4) = C_9^6 = 84$, $p(5) = C_{10}^6 = 210$ and $\sum_{k=1}^{5} p(k) = 330$.

Therefore the last six lucky numbers with 5 digits, from the largest to the smallest, are 70 000, 61 000, 60 100, 60 010, 60 001, 52 000. So The 325-th lucky number is 52 000, i.e. $a_{5n} = 52\,000$.

Part III Word Problems (Questions 13 to 15 carry 20 marks each.)

13 Given a sequence $\{a_n\}$ of numbers satisfying $a_0 = 1$, $a_{n+1} = \dfrac{7a_n + \sqrt{45a_n^2 - 36}}{2}$, $n \in \mathbf{N}$.

Prove that

(1) for each $n \in \mathbf{N}$, a_n is a positive integer.

(2) for each $n \in \mathbf{N}$, $a_n a_{n+1} - 1$ is a perfect square.

Proof (1) By assumption, $a_1 = 5$ and $\{a_n\}$ is strictly increasing with

$$2a_{n+1} - 7a_n = \sqrt{45a_n^2 - 36}.$$

Square both sides, and we get

$$a_{n+1}^2 - 7a_n a_{n+1} + a_n^2 + 9 = 0, \qquad \text{①}$$

$$a_n^2 - 7a_{n-1}a_n + a_{n-1}^2 + 9 = 0, \qquad \text{②}$$

$$\text{①} - \text{②}: a_{n+1} = 7a_n - a_{n-1}. \qquad \text{③}$$

It follows from $a_0 = 1$, $a_1 = 5$ and ③ that a_n is a positive integer for each $n \in \mathbf{N}$.

(2) From ①, we get $(a_{n+1} + a_n)^2 = 9(a_n a_{n+1} - 1)$,

so
$$a_{n+1} a_n - 1 = \left(\frac{a_{n+1} + a_n}{3}\right)^2.$$

By (1), a_n, a_{n+1} are positive integers and therefore $\dfrac{a_{n+1} + a_n}{3}$ is a rational number. Since $\left(\dfrac{a_{n+1} + a_n}{3}\right)^2 = a_{n+1} a_n - 1$ is a positive integer, so is $\dfrac{a_{n+1} + a_n}{3}$. Thus $a_{n+1} a_n - 1$ is the square of an integer.

⑫ Nine balls, numbered 1, 2, ⋯, 9, are put randomly at 9 equally spaced points on a circle, each point with a ball. Let S be the sum of the absolute values of the differences of the numbers of all two neighboring balls. Find the probability of S to be the minimum value. (Remark: If one arrangement of the balls is congruent to another after a rotation or a reflection, the two arrangements are regarded as the same).

Solution   9 balls with different numbers are placed at 9 equally spaced points on a circle, one point for one ball. This is equivalent to a circular arrangement of 9 distinct elements on a circle. Thus there are 8! arrangements. Considering the reflections, there are $\dfrac{8!}{2}$ essentially different arrangements.

Next, we calculate the number of arrangements, which make S the minimum. Along the circle there are two routes from 1 to 9, the major arc and the minor arc. For each of them, let x_1, x_2, ⋯, x_k be the numbers of the successive balls on the arc, then

$$|1 - x_1| + |x_1 - x_2| + \cdots + |x_k - 9|$$
$$\geqslant |(1 - x_1) + (x_1 - x_2) + \cdots + (x_k - 9)|$$
$$= |1 - 9| = 8.$$

The equality occurs if and only if $1 < x_1 < x_2 < \cdots < x_k < 9$, i.e.

the numbers of the balls on each route is increasing from 1 to 9. Therefore, $S_{min} = 2 \cdot 8 = 16$.

From the above analysis, when the numbers of the balls $\{1, x_1, x_2, \cdots, x_k, 9\}$ on each arc are fixed, the arrangement which gets the minimum value is uniquely determined. Divide the set of 7 balls $\{2, 3, \cdots, 8\}$ into two subsets, then the subset which contains less elements has $C_7^0 + C_7^1 + C_7^2 + C_7^3 = 2^6$ cases. Each case corresponds to a unique arrangement, which achieves the minimum value of S. Thus, the number of the arrangements when S takes the minimun value is 2^6 and the corresponding probability is $p = \dfrac{2^6}{\dfrac{8!}{2}} = \dfrac{1}{315}$.

⑮ Draw a tangent line of parabola $y = x^2$ at the point $A(1, 1)$. Suppose the line intersects the x-axis and y-axis at D and B respectively. Let point C be on the parabola and point E on AC such that $\dfrac{AE}{EC} = \lambda_1$. Let point F be on BC such that $\dfrac{BF}{FC} = \lambda_2$ and $\lambda_1 + \lambda_2 = 1$. Assume that CD intersects EF at point P. When point C moves along the parabola, find the equation of the trail of P.

Solution I   The slope of the tangent line passing through A is $y' = 2x \mid_{x=1} = 2$. So the equation of the tangent line AB is $y = 2x - 1$. Hence The coordinates of B and D are $B(0, -1)$, $D\left(\dfrac{1}{2}, 0\right)$. Thus D is the midpoint of line segment AB.

Consider $P(x, y)$, $C(x_0, x_0^2)$, $E(x_1, y_1)$, $F(x_2, y_2)$. Then by $\dfrac{AE}{EC} = \lambda_1$, we know $x_1 = \dfrac{1 + \lambda_1 x_0}{1 + \lambda_1}$, $y_1 = \dfrac{1 + \lambda_1 x_0^2}{1 + \lambda_1}$. From $\dfrac{BE}{FC} = \lambda_2$, we get $x_2 = \dfrac{\lambda_2 x_0}{1 + \lambda_2}$, $y_2 = \dfrac{-1 + \lambda_2 x_0^2}{1 + \lambda_2}$.

Therefore the equation of line EF is

$$\dfrac{y - \dfrac{1 + \lambda_1 x_0^2}{1 + \lambda_1}}{\dfrac{-1 + \lambda_2 x_0^2}{1 + \lambda_2} - \dfrac{1 + \lambda_1 x_0^2}{1 + \lambda_1}} = \dfrac{x - \dfrac{1 + \lambda_1 x_0}{1 + \lambda_1}}{\dfrac{\lambda_2 x_0}{1 + \lambda_2} - \dfrac{1 + \lambda_1 x_0}{1 + \lambda_1}}.$$

Simplifying it, we get

$$[(\lambda_2 - \lambda_1)x_0 - (1 + \lambda_2)]y$$
$$= [(\lambda_2 - \lambda_1)x_0^2 - 3]x + 1 + x_0 - \lambda_2 x_0^2. \tag{1}$$

When $x_0 \neq \dfrac{1}{2}$, the equation of line CD is

$$y = \frac{2x_0^2 x - x_0^2}{2x_0 - 1}. \tag{2}$$

From (1) and (2), we get $\begin{cases} x = \dfrac{x_0 + 1}{3}, \\ y = \dfrac{x_0}{3}. \end{cases}$

Eliminating x_0, we get the equation of the trail of point P as $y = \dfrac{1}{3}(3x - 1)^2$.

When $x_0 = \dfrac{1}{2}$, the equation of EF is $-\dfrac{3}{2}y = \left(\dfrac{1}{4}\lambda_2 - \dfrac{1}{4}\lambda_1 - 3\right)$ $x + \dfrac{3}{2} - \dfrac{1}{4}\lambda_2$, the equation of CD is $x = \dfrac{1}{2}$. Combining them, we conclude that $(x, y) = \left(\dfrac{1}{2}, \dfrac{1}{12}\right)$ is on the trail of P. Since C and A cannot be congruent, $x_0 \neq 1$, $x \neq \dfrac{2}{3}$.

Therefore the equation of the trail is $y = \dfrac{1}{3}(3x - 1)^2$, $x \neq \dfrac{2}{3}$.

Solution II From Solution I, the equation of AB is $y = 2x - 1$, $B(0, -1)$, $D\left(\dfrac{1}{2}, 0\right)$. Thus D is the midpoint of AB.

Set $\gamma = \dfrac{CD}{CP}$, $t_1 = \dfrac{CA}{CE} = 1 + \lambda_1$, $t_2 = \dfrac{CB}{CF} = 1 + \lambda_2$. Then $t_1 + t_2 = 3$.

Since AD is a median of $\triangle ABC$, $S_{\triangle CAB} = 2S_{\triangle CAD} = 2S_{\triangle CBD}$ where $S_\triangle$ denotes the area of $\triangle$. But

$$\frac{1}{t_1 t_2} = \frac{CE \cdot CF}{CA \cdot CB} = \frac{S_{\triangle CEF}}{S_{\triangle CAB}} = \frac{S_{\triangle CEP}}{2S_{\triangle CAD}} + \frac{S_{\triangle CFP}}{2S_{\triangle CED}}$$

$$= \frac{1}{2}\left(\frac{1}{t_1\gamma} + \frac{1}{t_2\gamma}\right) = \frac{t_1 + t_2}{2t_1 t_2 \gamma} = \frac{3}{2t_1 t_2 \gamma},$$

so $\gamma = \dfrac{3}{2}$ and P is the center of gravity for $\triangle ABC$.

Consider $P(x, y)$ and $C(x_0, x_0^2)$. Since C is different from A, $x_0 \neq 1$. Thus the coordinates of the center of gravity P are $x = \dfrac{0+1+x_0}{3} = \dfrac{1+x_0}{3}$, $x \neq \dfrac{2}{3}$, $y = \dfrac{-1+1+x_0^2}{3} = \dfrac{x_0^2}{3}$. Eliminating x_0, we get $y = \dfrac{1}{3}(3x-1)^2$. Thus the equation of the trail is $y = \dfrac{1}{3}(3x-1)^2$, $x \neq \dfrac{2}{3}$.

China Mathematical Competition (Extra Test)

2002 (Jilin)

1 As shown in the diagram, in $\triangle ABC$, $\angle A = 60°$, $AB > AC$, point O is a circumcenter and H is the intersection point of two altitudes BE and CF. Points M and N are on the line segments BH and HF respectively, and satisfy $BM = CN$. Determine the value of $\dfrac{MH + NH}{OH}$.

Solution We take $BK = CH$ on BE and join OB, OC and OK.

From the property of the circumcenter of a triangle, we know that $\angle BOC = 2\angle A = 120°$. From the property of the

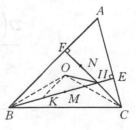

orthocenter of a triangle, we get $\angle BHC = 180° - \angle A = 120°$. So $\angle BOC = \angle BHC$. Then four points B, C, H and O are concyclic. Hence $\angle OBH = \angle OCH$.

In addition, $OB = OC$ and $BK = CH$. Therefore, $\triangle BOK \cong \triangle COH$. It follows that $\angle BOK = \angle COH$, and $OK = OH$.

So,
$$\angle KOH = \angle BOC = 120°,$$
$$\angle OKH = \angle OHK = 30°.$$

In $\triangle OKH$, by the sine rule, we get $KH = \sqrt{3}OH$. In view of $BM = CN$ and $BK = CH$, we get $KM = NH$, and
$$MH + NH = MH + KM = KH = \sqrt{3}OH.$$

Therefore,
$$\frac{MH + NH}{OH} = \sqrt{3}.$$

⚫ There are real numbers a, b and c and a positive number λ such that $f(x) = x^3 + ax^2 + bx + c$ has three real roots x_1, x_2 and x_3 satisfying

(1) $x_2 - x_1 = \lambda$,

(2) $x_3 > \dfrac{1}{2}(x_1 + x_2)$.

Find the maximum value of $\dfrac{2a^3 + 27c - 9ab}{\lambda^3}$.

Solution   Let $S = \dfrac{2a^3 + 27c - 9ab}{\lambda^3}$, then

$$S = \frac{27\left(\dfrac{2}{27}a^3 - \dfrac{1}{3}ab + c\right)}{\lambda^3} = \frac{27f\left(-\dfrac{1}{3}a\right)}{\lambda^3}$$

$$= \frac{27\left(-\dfrac{1}{3}a - x_1\right)\left(-\dfrac{1}{3}a - x_2\right)\left(-\dfrac{1}{3}a - x_3\right)}{(x_2 - x_1)^3}$$

$$= \frac{-27\left(x_1 + \dfrac{1}{3}a\right)\left(x_2 + \dfrac{1}{3}a\right)\left(x_3 + \dfrac{1}{3}a\right)}{(x_2 - x_1)^3}.$$

Write $u_i = x_i + \dfrac{a}{3}\,(i=1,\,2,\,3)$, then $u_2 - u_1 = x_2 - x_1 = \lambda$, $u_3 >$

$\dfrac{1}{2}(u_1 + u_2)$, and $u_1 + u_2 + u_3 = x_1 + x_2 + x_3 + a = 0$. So, u_1, u_2 and

u_3 satisfy the corresponding conditions too, and

$$S = \frac{-27 u_1 u_2 u_3}{(u_2 - u_1)^3}.$$

By $u_3 = -(u_1 + u_2) > \dfrac{u_1 + u_2}{2}$, we get $u_1 + u_2 < 0$. Consequently,

at least one of u_1 and u_2 should be less than 0. We may assume $u_1 < 0$.

If $u_2 < 0$, then $S < 0$.

If $u_2 > 0$, we suppose further

$$v_1 = \frac{-u_1}{u_2 - u_1}, \quad v = v_2 = \frac{u_2}{u_2 - u_1}.$$

Then $v_1 + v_2 = 1$, v_1 and v_2 are both greater than 0 and $v_1 - v_2 =$

$\dfrac{u_3}{u_2 - u_1} > 0$. So it follows that

$$S = 27 v_1 v_2 (v_1 - v_2) = 27 v (1 - v)(1 - 2v)$$

$$= 27 \sqrt{(v - v^2)^2 (1 - 2v)^2}$$

$$= 27 \sqrt{2(v - v^2)(v - v^2)\left(\frac{1}{2} - 2v + 2v^2\right)}$$

$$\leqslant 27 \sqrt{2 \times \left(\frac{1}{6}\right)^3}$$

$$= \frac{3}{2} \sqrt{3}.$$

The equality holds when $v = \dfrac{1}{2}\left[1 - \dfrac{\sqrt{3}}{3}\right]$. The corresponding cubic

equation is $x^3 - \dfrac{1}{2} x + \dfrac{\sqrt{3}}{18} = 0$ and $\lambda = 1$ $\left(x_1 = -\dfrac{1}{2}\left(1 + \dfrac{\sqrt{3}}{3}\right),\ x_2 = \right.$

$\frac{1}{2}\left(1-\frac{\sqrt{3}}{3}\right)$ and $x_3 = \frac{\sqrt{3}}{3}$.

Consequently, the maximal value is $\frac{3}{2}\sqrt{3}$.

Before The World Cup tournament, the football coach of F country will let seven players, A_1, A_2, $\cdots$, A_7, join three training matches (90 minutes each) in order to assess them. Suppose, at any moment during a match, one and only one of them enters the field, and the total time (which is measured in minutes) on the field for each one of A_1, A_2, A_3 and A_4 is divisible by 7 and the total time for each of A_5, A_6 and A_7 is divisible by 13. If there is no restriction about the number of times of substitution of players during each match, then how many possible cases are there within the total time for every player on the field?

Solution Suppose that x_i ($i = 1, 2, \cdots, 7$) minutes is the time for i-th player on the field. Now, the problem is to find the number of solution groups of positive integers for the following equation:

$$x_1 + x_2 + \cdots + x_7 = 270 \qquad (1)$$

when the conditions $7 \mid x_i$ ($i = 1, 2, 3, 4$) and $13 \mid x_j$ ($j = 5, 6, 7$) are satisfied.

Suppose $x_1 + x_2 + x_3 + x_4 = 7m$ and $x_5 + x_6 + x_7 = 13n$. Then

$$7m + 13n = 270,$$

and $m, n \in \mathbf{N}_+$, $m \geqslant 4$ and $n \geqslant 3$.

It is easy to find the positive integer solutions (m, n) to be

$$(m, n) = (33, 3), (20, 10), (7, 17)$$

which satisfy the conditions above.

When $(m, n) = (33, 3)$, $x_5 = x_6 = x_7 = 13$. Let $x_i = 7y_i$ ($i = 1, 2, 3, 4$), then

$$y_1 + y_2 + y_3 + y_4 = 33.$$

We get $C_{33-1}^{4-1} = C_{32}^3 = 4\,960$ solution groups of positive integers (y_1, y_2, y_3, y_4) and in this case, we have 4 960 solution groups of positive integers satisfying the conditions.

When $(m, n) = (20, 10)$, let $x_i = 7y_i (i = 1, 2, 3, 4)$ and $x_j = 13y_j$ $(j = 5, 6, 7)$. Hence

$$y_1 + y_2 + y_3 + y_4 = 20 \text{ and } y_5 + y_6 + y_7 = 10.$$

In this case, we have $C_{19}^3 \times C_9^2 = 34\,884$ solution groups of positive integers satisfying the conditions.

When $(m, n) = (7, 17)$, set $x_i = 7y_i (i = 1, 2, 3, 4)$ and $x_j = 13y_j (j = 5, 6, 7)$. Hence

$$y_1 + y_2 + y_3 + y_4 = 7 \text{ and } y_5 + y_6 + y_7 = 17.$$

In this case, we have $C_6^3 \times C_{16}^3 = 2\,400$ solution groups of positive integers satisfying the conditions.

Consequently, for (1), there are

$$4\,960 + 34\,884 + 2\,400 = 42\,244$$

solution groups of positive integers satisfying the conditions.

2003 (Shaanxi)

From point P outside a circle draw two tangents to the circle touching at points A and B. Draw a secant line intersecting the circle at points C and D, with C between P and D. Choose point Q on the chord CD such that $\angle DAQ = \angle PBC$. Prove that $\angle DBQ = \angle PAC$.

Solution Using $\angle DAB = \angle DCB$, $\angle DAB = \angle DAQ + \angle QAB$, $\angle DCB = \angle PBC + \angle BPQ$, and $\angle DAQ = \angle PBC$, we get

$\angle QAB = \angle BPQ$, so points P, A, Q, B share a common circle. Then $\angle BQP = \angle PAB$, that is, $\angle DBQ + \angle CDB = \angle PAC + \angle CAB$. Since $\angle CDB = \angle CAB$, so $\angle DBQ = \angle PAC$.

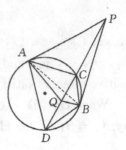

Remark The condition that PA and PB are tangents to the circle is not necessary. Making use of this condition intentionally may lead to further complication.

Let the three sides of a triangle be integers l, m, n, respectively, satisfying $l > m > n$ and $\left\{\dfrac{3^l}{10^4}\right\} = \left\{\dfrac{3^m}{10^4}\right\} = \left\{\dfrac{3^n}{10^4}\right\}$, where $\{x\} = x - [x]$ and $[x]$ denotes the integral part of the number x. Find the minimum perimeter of such a triangle.

Solution Since

$$\frac{3^l}{10^4} - \left[\frac{3^l}{10^4}\right] = \frac{3^m}{10^4} - \left[\frac{3^m}{10^4}\right] = \frac{3^n}{10^4} - \left[\frac{3^n}{10^4}\right],$$

we have

$$3^l \equiv 3^m \equiv 3^n \,(\mathrm{mod}\,10^4)$$

$$\Leftrightarrow \begin{cases} 3^l \equiv 3^m \equiv 3^n \,(\mathrm{mod}\,2^4), & (1) \\ 3^l \equiv 3^m \equiv 3^n \,(\mathrm{mod}\,5^4). & (2) \end{cases}$$

As $(3, 2) = 1$, we then have from (1) $3^{l-n} \equiv 3^{m-n} \equiv 1 \,(\mathrm{mod}\,2^4)$.

Let u be the minimum positive integer satisfying $3^u \equiv 1 \,(\mathrm{mod}\,2^4)$. Then for every positive integer v satisfying $3^v \equiv 1 \,(\mathrm{mod}\,2^4)$, we must have $u \mid v$. Otherwise, if $u \nmid v$, then using division with a remainder we could get two non-negative integers a and b satisfying $v = au + b$ with $0 < b < u$. Then $3^b \equiv 3^{au+b} \equiv 3^v \equiv 1 \,(\mathrm{mod}\,2^4)$, contradicting the definition of u. Therefore $u \mid v$.

Notice that

$$3 \equiv 3 \,(\mathrm{mod}\,2^4),\ 3^2 \equiv 9 \,(\mathrm{mod}\,2^4),$$

$$3^3 \equiv 27 \equiv 11 (\mathrm{mod}\, 2^4), \ 3^4 \equiv 1 (\mathrm{mod}\, 2^4),$$

then $u = 4$. We may assume that $m - n = 4k$, where k is a positive integer.

In the same way, we get from (2) $3^{m-n} \equiv 1 (\mathrm{mod}\, 5^4)$, that is, $3^{4k} \equiv 1 (\mathrm{mod}\, 5^4)$.

Now, we are going to find number k. As $3^{4k} - 1 = (1 + 5 \times 2^4)^k - 1 = 0\,\mathrm{mod}\, 5^4$, i.e.

$$5k \times 2^4 + \frac{k(k-1)}{2} \times 5^2 \times 2^8 + \frac{k(k-1)(k-2)}{6} \times 5^3 \times 2^{12}$$

$$\equiv 5k + 5^2 k[3 + (k-1) \times 2^7] + \frac{k(k-1)(k-2)}{3} \times 5^3 \times 2^{11}$$

$$\equiv 0 (\mathrm{mod}\, 5^4),$$

so $k = 5t$. Substituting in the above expression, we get

$$t + 5t[3 + (5t - 1) \times 2^7] \equiv 0 (\mathrm{mod}\, 5^2).$$

Then $k = 5t = 5^3 s$, and $m - n = 500s$, where s is a positive integer.

In the same way, we get $l - n = 500r$, where r is a positive integer and $r > s$ since $l > m > n$.

So the three sides of the required triangle are $l = 500r + n$, $m = 500s + n$ and n, respectively, satisfying $n > l - n = 500(r - s)$. When $s = 1$, $r = 2$ the perimeter reaches the minimum which equals $(1\,000 + 501) + (500 + 501) + 501 = 3\,003$.

Remark The key to solve the problem is to find the least positive integer u satisfying $3^u \equiv 1 (\mathrm{mod}\, 10^4)$, which, in number theory, is called the order of 3 with respect to modulus 10^4.

Let a space figure consist of n vertices and l lines connecting these vertices, with $n = q^2 + q + 1$, $l \geqslant q^2 (q+1)^2 + 1$, $q \geqslant 2$, $q \in \mathbf{N}$. Suppose the figure satisfies the following conditions: every four vertices are non-coplanar, every vertex is connected by at least one line, and there is a vertex which is connected by at least $q + 2$ lines. Prove that there exists a space quadrilateral in the figure, i.e. a quadrilateral with four vertices A, B, C, D and four lines AB, BC, CD, DA in the figure.

Solution Let $V = \{A_0, A_1, A_2, \cdots, A_{n-1}\}$ be the set of all the n vertices, B_i the set of all vertices adjacent to vertex A_i (i.e. connected with A_i by a line in the figure), and the number of the elements in B_i denoted by $|B_i| = b_i$. Obviously,

$$\sum_{i=1}^{n-1} b_i = 2l \text{ and } b_i \leqslant (n-1), \ i = 0, 1, 2, \cdots, n-1.$$

If there exists i such that $b_i = n-1$, without losing generality, we assume that $i = n-1$ and $q+2 \leqslant n-1$, and then we have

$$l' = (n-1) + \left[\frac{n-1}{2}\right] + 1 \leqslant \frac{1}{2}q(q+1)^2 + 1 \leqslant l.$$

That means there exists $b_j \geqslant 2$ $(0 \leqslant j \leqslant n-2)$, so there must be a space quadrilateral including A_j and A_{n-1} as its vertices in the figure.

Then we consider the case when $b_i < n-1$, $i = 0, 1, 2, \cdots, n-1$, and we may assume that $q+2 \leqslant b_0$.

We will give the proof by reduction to absurdity. If there is no such a quadrilateral in the figure, B_i and B_j share no vertex-pair when $i \neq j$, then

$$|B_i \cap B_j| \leqslant 1 \quad (0 \leqslant i < j \leqslant n-1).$$

So

$$|B_i \cap \overline{B}_0| \geqslant b_i - 1, \ i = 1, 2, \cdots, n-1.$$

Then we have the number of vertex-pairs in $V \cap \overline{B}_0 = C_{n-b_0}^2$

$$\geqslant \sum_{i=1}^{n-1} (\text{number of vertex-pairs in} |B_i \cap \overline{B}_0|) = \sum_{i=1}^{n-1} C_{|B_i \cap \overline{B}_0|}^2$$

$$\geqslant \sum_{i=1}^{n-1} C_{|b_i-1|}^2 \ (\text{Let } C_{|b_i-1|}^2 = 0, \text{ when } b_i = 0 \text{ or } 1)$$

$$= \frac{1}{2} \sum_{i=1}^{n-1} (b_i^2 - 3b_i + 2)$$

$$\geq \frac{1}{2}\left[\frac{(\sum_{i=1}^{n-1} b_i)^2}{n-1} - 3\left(\sum_{i=1}^{n-1} b_i\right) + 2(n-1)\right]$$

$$= \frac{1}{2(n-1)}\left[\sum_{i=1}^{n-1} b_i - (n-1)\right]\left[\sum_{i=1}^{n-1} b_i - 2(n-1)\right]$$

$$= \frac{1}{2(n-1)}(2l - b_0 - n + 1)(2l - b_0 - 2n + 2)$$

$$\geq \frac{1}{2(n-1)}(nq - q + 2 - b_0)(nq - q + 3 - b_0).$$

So $(n-1)(n-b_0)(n-b_0-1) \geq (nq-q+2-b_0)(nq-q-n+3-b_0)$. That is

$$q(q+1)(n-b_0)(n-b_0-1)$$

$$\geq (nq-q+2-b_0)(nq-q-n+3-b_0). \tag{1}$$

On the other hand,

$$(nq-q-n+3-b_0) - q(n-b_0-1)$$

$$= (q-1)b_0 - n + 3 \geq (q-1)(q+2) - n + 3 = 0, \tag{2}$$

and

$$(nq-q+2-b_0) - (q+1)(n-b_0) = qb_0 - q - n + 2$$

$$\geq q(q+2) - q - n + 2 = 1 > 0. \tag{3}$$

As $(n-b_0)(q+1)$, $(n-b_0-1)q$ are positive numbers, we then get from (2) and (3),

$$(nq-q+2-b_0)(nq-q-n+3-b_0)$$

$$> q(q+1)(n-b_0)(n-b_0-1),$$

which contradicts to (1). This completes the proof.

Remark The question here derives from a research topic in graph theory, which investigates under what conditions a graph with n vertices and l edges contains a quadrilateral.

2004 (Hainan)

In an acute triangle ABC, point H is the intersection point of altitude CE to AB and altitude BD to AC. A circle with DE as its diameter intersects AB and AC at points F and G, respectively. FG and AH intersect at point K. If $BC = 25$, $BD = 20$, and $BE = 7$, find the length of AK.

Solution We know that $\angle ADB = \angle AEC = 90°$, therefore

$$\triangle ADB \backsim \triangle AEC,$$

and

$$\frac{AD}{AE} = \frac{BD}{CE} = \frac{AB}{AC}. \tag{1}$$

But $BC = 25$, $BD = 20$, and $BE = 7$, so $CD = 15$, and $CE = 24$. From (1), we obtain

$$\begin{cases} \dfrac{AD}{AE} = \dfrac{5}{6}, \\ \dfrac{AE+7}{AD+15} = \dfrac{5}{6}, \end{cases}$$

and the solution is

$$\begin{cases} AD = 15, \\ AE = 18. \end{cases}$$

Thus, point D is the midpoint of the hypotenuse AC of Rt$\triangle AEC$, and

$$DE = \frac{1}{2}AC = 15.$$

Draw line DF. Since point F is on the

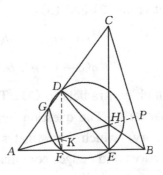

circle with DE as its diameter, $\angle DFE = 90°$, we have

$$AF = \frac{1}{2}AE = 9.$$

Since four points G, F, E and D are concyclic, and four points D, E, B and C are concyclic too, we get

$$\angle AFG = \angle ADE = \angle ABC.$$

Thus $GF \parallel CB$. Extend line AH to intersect BC at point P, then

$$\frac{AK}{AP} = \frac{AF}{AB}. \tag{2}$$

Since H is the orthocenter of $\triangle ABC$, $AP \perp BC$. From $BA = BC$ we have

$$AP = CE = 24.$$

Due to (2), we get

$$AK = \frac{AF \cdot AP}{AB} = \frac{9 \times 24}{25} = 8.64.$$

In a planar rectangular coordinate system, a sequence of points $\{A_n\}$ on the positive half of the y-axis and a sequence of points $\{B_n\}$ on the curve $y = \sqrt{2x}$ $(x \geqslant 0)$ satisfy the condition $|OA_n| = |OB_n| = \frac{1}{n}$. The x-intercept of line segment A_nB_n is a_n, and the x-coordinate of point B_n is b_n, $n \in \mathbf{N}$. Prove that

(1) $a_n > a_{n+1} > 4$, $n \in \mathbf{N}$;

(2) There is $n_0 \in \mathbf{N}$, such that for any $n > n_0$, $\dfrac{b_2}{b_1} + \dfrac{b_3}{b_2} + \cdots + \dfrac{b_n}{b_{n-1}}$

$+ \dfrac{b_{n+1}}{b_n} < n - 2\,004.$

Proof (1) According to the stated conditions we have $A_n\left(0, \dfrac{1}{n}\right)$, and $B_n\left(b_n, \sqrt{2b_n}\right)$ $(b_n > 0)$. From $\{OB_n\} = \dfrac{1}{n}$ we get

$$b_n^2 + 2b_n = \left(\frac{1}{n}\right)^2.$$

Thus

$$b_n = \sqrt{\left(\frac{1}{n}\right)^2 + 1} - 1, \ n \in \mathbf{N}.$$

On the other hand, the x-intercept a_n of line segment $A_n B_n$ satisfies the following equation

$$(a_n - 0)\left(\sqrt{2b_n} - \frac{1}{n}\right) = \left(0 - \frac{1}{n}\right)(b_n - 0).$$

Hence,

$$a_n = \frac{b_n}{1 - n\sqrt{2b_n}}, \ n \in \mathbf{N}.$$

Since $2n^2 b_n = 1 - n^2 b_n^2 > 0$, we have $b_n + 2 = \dfrac{1}{n^2 b_n}$ and

$$a_n = \frac{b_n\left(1 + n\sqrt{2b_n}\right)}{1 - 2n^2 b_n} = \frac{b_n\left(1 + n\sqrt{2b_n}\right)}{1 - (1 - n^2 b_n^2)}$$

$$= \frac{1}{n^2 b_n} + \frac{\sqrt{2}}{\sqrt{n^2 b_n}} = b_n + 2 + \sqrt{2(b_n + 2)}.$$

Thus, $a_n = \sqrt{\left(\dfrac{1}{n}\right)^2 + 1} + 1 + \sqrt{2\sqrt{\left(\dfrac{1}{n}\right)^2 + 1} + 2}$. Since $\dfrac{1}{n} > \dfrac{1}{n+1} > 0$,

$$a_n > a_{n+1} > 4 \text{ for any } n \in \mathbf{N}.$$

(2) Let $c_n = 1 - \dfrac{b_{n+1}}{b_n}$ for $n \in \mathbf{N}$, then

$$c_n = \frac{\sqrt{\left(\frac{1}{n}\right)^2 + 1} - \sqrt{\left(\frac{1}{n+1}\right)^2 + 1}}{\sqrt{\left(\frac{1}{n}\right)^2 + 1} - 1}$$

$$= n^2 \left[\frac{1}{n^2} - \frac{1}{(n+1)^2}\right] \cdot \frac{\sqrt{\left(\frac{1}{n}\right)^2 + 1} + 1}{\sqrt{\left(\frac{1}{n}\right)^2 + 1} + \sqrt{\left(\frac{1}{n+1}\right)^2 + 1}}$$

$$> \frac{2n+1}{(n+1)^2} \left[\frac{1}{2} + \frac{1}{2\sqrt{\left(\frac{1}{n}\right)^2 + 1}}\right] > \frac{2n+1}{2(n+1)^2}.$$

Since $(2n+1)(n+2) - 2(n+1)^2 = n > 0$, we obtain

$$c_n > \frac{1}{n+2}, \quad n \in \mathbf{N}.$$

Let $S_n = c_1 + c_2 + \cdots + c_n$, $n \in \mathbf{N}$. If $n = 2^k - 2 > 1$ $(k \in \mathbf{N})$, then

$$S_n > \frac{1}{3} + \frac{1}{4} + \cdots + \frac{1}{2^k - 1} + \frac{1}{2^k}$$

$$= \left(\frac{1}{3} + \frac{1}{4}\right) + \left[\frac{1}{2^2 + 1} + \cdots + \frac{1}{2^3}\right] + \cdots + \left[\frac{1}{2^{k-1} + 1} + \cdots + \frac{1}{2^k}\right]$$

$$> 2 \cdot \frac{1}{2^2} + 2^2 \cdot \frac{1}{2^3} + \cdots + 2^{k-1} \cdot \frac{1}{2^k} = \frac{k-1}{2}.$$

Therefore, if we put $n_0 = 2^{4\,009} - 2$, then for any $n > n_0$ we have

$$\left[1 - \frac{b_2}{b_1}\right] + \left[1 - \frac{b_3}{b_2}\right] + \cdots + \left[1 - \frac{b_{n+1}}{b_n}\right] = S_n > S_{n_0} > \frac{4\,009 - 1}{2}$$

$$= 2\,004.$$

Consequently,

$$\frac{b_2}{b_1} + \frac{b_3}{b_2} + \cdots + \frac{b_n}{b_{n-1}} + \frac{b_{n+1}}{b_n} < n - 2\,004, \quad n > n_0.$$

Remark   To prove (1), it is mainly required to determine the expression of a_n which has many different forms, its monotonicity can be observed directly from the expression of a_n. So some students pointed out that $a_n > a_{n+1}$ is obvious after writing down the expression of a_n.

About proving (2) one can refer to the test paper of the 19th Mathematical Olympiad of Soviet Union (1985).

⬤ For integer $n \geqslant 4$, find the minimal integer $f(n)$, such that for any positive integer m, in any subset with $f(n)$ elements of the set $\{m, m+1, \cdots, m+n-1\}$ there are at least 3 mutually prime elements.

Solution Ⅰ   When $n \geqslant 4$, we consider the set

$$M = \{m, m+1, m+2, \cdots, m+n-1\}.$$

If $2 \mid m$, then $m+1, m+2, m+3$ are mutually prime;
If $2 \nmid m$, then $m, m+1, m+2$ are mutually prime.

Therefore, in every n-element subset of M, there are at least 3 mutually prime elements. Hence there exists $f(n)$ and

$$f(n) \leqslant n.$$

Let $T_n = \{t \mid t \leqslant n+1 \text{ and } 2 \mid t \text{ or } 3 \mid t\}$, then T_n is a subset of $\{2, 3, \cdots, n+1\}$. But any 3 elements in T_n are not mutually prime, thus $f(n) \geqslant \mid T_n \mid + 1$.

By the inclusion and exclusion principle, we have

$$\mid T_n \mid = \left[\frac{n+1}{2}\right] + \left[\frac{n+1}{3}\right] - \left[\frac{n+1}{6}\right].$$

Thus

$$f(n) \geqslant \left[\frac{n+1}{2}\right] + \left[\frac{n+1}{3}\right] - \left[\frac{n+1}{6}\right] + 1. \tag{1}$$

Therefore

$$f(4) \geqslant 4, \ f(5) \geqslant 5, \ f(6) \geqslant 5,$$

$$f(7) \geqslant 6, \ f(8) \geqslant 7, \ f(9) \geqslant 8.$$

Now we prove that $f(6) = 5$.

Let x_1, x_2, x_3, x_4, x_5 be 5 numbers in $\{m, m+1, \cdots, m+5\}$. If among these 5 numbers there are 3 odds, then they are mutually prime. If there are 2 odds among these 5 numbers, then the other three numbers are even, say x_1, x_2, x_3, and the 2 odds are x_4, x_5. When $1 \leqslant i < j \leqslant 3$, $|x_i - x_j| \in \{2, 4\}$. Thus among x_1, x_2, x_3 there is at most one which is divisible by 3, and at most one which is divisible by 5. Therefore, there is at least one which is neither divisible by 3 nor by 5, say, $3 \nmid x_3$ and $5 \nmid x_3$. Then x_3, x_4, x_5 are mutually prime. This is to say, among these 5 numbers there are 3 elements which are mutually prime, i.e. $f(6) = 5$.

On the other hand, $\{m, m+1, \cdots, m+n\} = \{m, m+1, \cdots, m+n-1\} \cup \{m+n\}$ implies that

$$f(n+1) \leqslant f(n) + 1.$$

Since $f(6) = 5$, we have

$$f(4) = 4, \ f(5) = 5, \ f(7) = 6, \ f(8) = 7, \ f(9) = 8.$$

Thus when $4 \leqslant n \leqslant 9$,

$$f(n) = \left[\frac{n+1}{2}\right] + \left[\frac{n+1}{3}\right] - \left[\frac{n+1}{6}\right] + 1. \tag{2}$$

In the following we will prove that (2) holds for all n by mathematical induction.

Suppose that equation (2) holds for all $n \leqslant k$ $(k \geqslant 9)$. In the case when $n = k+1$, since

$$\{m, m+1, \cdots, m+k\} = \{m, m+1, \cdots, m+k-6\} \cup$$

$$\{m+k-5, m+k-4, m+k-3, m+k-2, m+k-1, m+k\},$$

equation (2) holds for $n = 6$, $n = k-5$, we have

$$f(k+1) \leqslant f(k-5) + f(6) - 1$$

$$= \left[\frac{k+2}{2}\right] + \left[\frac{k+2}{3}\right] - \left[\frac{k+2}{6}\right] + 1. \qquad (3)$$

By (1) and (3) we obtain that equation (2) holds for $n = k+1$. Consequently, for any $n \geqslant 4$, we have

$$f(n) = \left[\frac{n+1}{2}\right] + \left[\frac{n+1}{3}\right] - \left[\frac{n+1}{6}\right] + 1.$$

Solution II   At first, we verify that equations $f(4) = 4$, $f(5) = 5$, $f(6) = 5$ hold.

When $n = 4$, consider $\{m, m+1, m+2, m+3\}$. If m is odd, then m, $m+1$, $m+2$ are mutually prime. If m is even, then $m+1$, $m+2$, $m+3$ are mutually prime. Thus $f(4) \leqslant 4$. But from the set $\{m, m+1, m+2, m+3\}$ choose a 3-element subset consisting of two evens and one odd we know these three numbers are not mutually prime, which implies that $f(4) = 4$.

When $n = 5$, consider $\{m, m+1, m+2, m+3, m+4\}$. If m is even, then m, $m+2$, $m+4$ are all even. Then any 3 numbers from the 4-element subset $\{m, m+1, m+2, m+4\}$ are not mutually prime, so $f(5) > 4$. But from the 5-element universal set we can find out 3 numbers which are mutually prime. Thus $f(5) = 5$.

When $n = 6$, the elements of the set $\{m, m+1, m+2, m+3, m+4, m+5\}$ are 3 odd and 3 even. If we choose a 4-element subset consisting of 3 evens and 1 odd, then among the subset there are no 3 numbers which are mutually prime. Thus $f(6) > 4$. Consider the 5-element subset, if among these 5 elements there are 3 odds, then these 3 numbers must be mutually prime. If these 5 elements are 3 evens and 2 odds, then among these 3 evens, there is at most one number which is divisible by 3, and there is at most one number which is divisible by 5. Hence among these 3 evens there is one number which is neither divisible by 3 nor 5. The even number and the other 2 odds are mutually prime, which implies that $f(6) = 5$.

When $n > 6$, set $T_n = \{t \mid t \leqslant n+1 \text{ and } 2 \mid t \text{ or } 3 \mid t\}$. Then T_n is a subset of $\{2, 3, \cdots, n+1\}$, and any 3 elements in T_n are not mutually prime. So

$$f(n) \geqslant |T_n| + 1 = \left[\frac{n+1}{2}\right] + \left[\frac{n+1}{3}\right] - \left[\frac{n+1}{6}\right] + 1.$$

Assume $n = 6k + r$, $k \geqslant 1$, $r = 0, 1, 2, 3, 4, 5$, then

$$\left[\frac{n+1}{2}\right] + \left[\frac{n+1}{3}\right] - \left[\frac{n+1}{6}\right] + 1$$

$$= 4k + \left[\frac{r+1}{2}\right] + \left[\frac{r+1}{3}\right] - \left[\frac{r+1}{6}\right] + 1.$$

It is easy to verity that

$$\left[\frac{r+1}{2}\right] + \left[\frac{r+1}{3}\right] - \left[\frac{r+1}{6}\right] + 1 = \begin{cases} r, & r = 0, 1, 2, 3, \\ r-1, & r = 4, 5. \end{cases}$$

If $r = 0, 1, 2, 3$, we can divide $n = 6k + r$ numbers into k groups:

$$\{m, m+1, \cdots, m+5\}, \{m+6, m+7, \cdots, m+11\}, \cdots,$$
$$\{m+6(k-1), m+6k-5, \cdots, m+6k-1\},$$

and left with r numbers $m + 6k, \cdots, m + 6k + r - 1$. Among $4k + r + 1$ numbers, there are at least $4k + 1$ numbers which are contained in the above k groups. Thus there is at least one group which contains 5 numbers. Since $f(6) = 5$, there are 3 numbers which are mutually prime.

If $r = 4, 5$, we can prove similarly that there are 3 numbers which are mutually prime. Hence, $f(n) = \left[\frac{n+1}{2}\right] + \left[\frac{n+1}{3}\right] - \left[\frac{n+1}{6}\right] + 1.$

2005 (Jiangxi)

❶ In $\triangle ABC$, $AB > AC$, l is a tangent line of the circumscribed circle of $\triangle ABC$, passing through A. The circle, centered at A with radius AC, intersects AB at D, and line l at E, F (see the diagram). Prove that lines DE, DF pass through the incenter and a escenter of $\triangle ABC$ respectively.

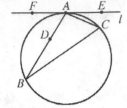

(Remark: The circle which is tangent to one side of a triangle and two other extended sides is called an escribed circle. The center of an escribed circle is called an escenter).

Proof First, prove that DE passes through the incenter of $\triangle ABC$. In fact, if we join DE, DC and draw the bisector of $\angle BAC$, the bisector intersects DE, DC at I, G respectively. Join IC. From $AD = AC$, we get $AG \perp DC$ and $ID = IC$.

Since D, C, E lie on a circle with center A, $\angle IAC = \dfrac{1}{2}\angle DAC = \angle IEC$. So A, I, C, E are on the same circle, and $\angle CIE = \angle CAE = \angle ABC$. But $\angle CIE = 2\angle ICD$, thus $\angle ICD = \dfrac{1}{2}\angle ABC$.

We get $\angle AIC = \angle IGC + \angle ICG$

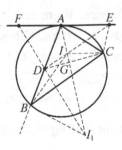

$$= 90° + \dfrac{1}{2}\angle ABC,$$

and $\qquad \angle ACI = \dfrac{1}{2}\angle ACB,$

so I is the incenter of $\triangle ABC$.

Secondly, DF passes through an escenter of $\triangle ABC$. In fact, the extension of FD intersects the bisector of the exterior angle of $\angle ABC$ at I_1. Join II_1, BI_1, BI. By (1), I is the incenter, we get $\angle IBI_1 = 90° = \angle EDI_1$. So D, B, I_1, I are on the same circle.

In view of $\qquad \angle IBI_1 = \angle BDI_1 = 90° - \angle ADI$

$$= \left(\dfrac{1}{2}\angle BAC + \angle ADG\right) - \angle ADI$$

$$= \frac{1}{2}\angle BAC + \angle IDG$$

so A, I, I_1 are on the same line. Furthermore, I_1 is the escenter of $\triangle ABC$ outside BC.

Assume that positive numbers a, b, c, x, y, z satisfy $cy + bz = a$; $az + cx = b$ and $bx + ay = c$. Find the minimum value of the function $f(x, y, z) = \dfrac{x^2}{1+x} + \dfrac{y^2}{1+y} + \dfrac{z^2}{1+z}$.

Solution By assumption, $b(az + cx - b) + c(bx + ay - c) - a(cy + bz - a) = 0$, i.e. $2bcx + a^2 - b^2 - c^2 = 0$, we get $x = \dfrac{b^2 + c^2 - a^2}{2bc}$. For

the similar reason, $y = \dfrac{a^2 + c^2 - b^2}{2ac}$ and $z = \dfrac{a^2 + b^2 - c^2}{2ab}$.

Since a, b, c, x, y, z are positive, by the above three expressions, we know $b^2 + c^2 > a^2$, $a^2 + c^2 > b^2$ and $a^2 + b^2 > c^2$. Thus there is an acute triangle ABC with the lengths of its sides a, b, c. So $x = \cos A$, $y = \cos B$ and $z = \cos C$. The problem is now changed to finding the minimum value of the function

$$f(\cos A, \cos B, \cos C) = \frac{\cos^2 A}{1 + \cos A} + \frac{\cos^2 B}{1 + \cos B} + \frac{\cos^2 C}{1 + \cos C}.$$

Set $u = \cot A$, $v = \cot B$, $w = \cot C$, then u, v, $w \in \mathbf{R}^+$, $uv + vw + wu = 1$, $u^2 + 1 = (u+v)(u+w)$, $v^2 + 1 = (u+v)(v+w)$ and $w^2 + 1 = (u+w)(v+w)$.

We get $\dfrac{\cos^2 A}{1 + \cos A} = \dfrac{\dfrac{u^2}{u^2 + 1}}{1 + \dfrac{u}{\sqrt{u^2 + 1}}} = \dfrac{u^2}{\sqrt{u^2 + 1}(\sqrt{u^2 + 1} + u)}$

$$= \frac{u^2(\sqrt{u^2 + 1} - u)}{\sqrt{u^2 + 1}} = u^2 - \frac{u^3}{\sqrt{u^2 + 1}}$$

$$= u^2 - \frac{u^3}{\sqrt{(u+v)(u+w)}}$$

$$\geqslant u^2 - \frac{u^3}{2}\left(\frac{1}{u+v} + \frac{1}{u+w}\right).$$

By a similar argument, $\dfrac{\cos^2 B}{1+\cos B} \geqslant v^2 - \dfrac{v^3}{2}\left(\dfrac{1}{u+v} + \dfrac{1}{v+w}\right)$

and $\dfrac{\cos^2 C}{1+\cos C} \geqslant w^2 - \dfrac{w^3}{2}\left(\dfrac{1}{u+w} + \dfrac{1}{v+w}\right).$

Hence $f \geqslant u^2 + v^2 + w^2 - \dfrac{1}{2}\left(\dfrac{u^3+v^3}{u+v} + \dfrac{w^3+v^3}{w+v} + \dfrac{u^3+w^3}{u+w}\right)$

$$= u^2 + v^2 + w^2 - \frac{1}{2}\Big[(u^2 - uv + v^2)$$

$$+ (v^2 - vw + w^2) + (u^2 - uw^2 + w^2)\Big]$$

$$= \frac{1}{2}(uv + vw + uw) = \frac{1}{2},$$

the equality sign is valid if and only if $u = v = w$, i.e. $a = b = c$, $x = y = z = \dfrac{1}{2}$,

so $[f(x, y, z)]_{\min} = \dfrac{1}{2}.$

⬡ For each positive integer, define a function

$$f(n) = \begin{cases} 0, & \text{if } n \text{ is the square of an integer,} \\[2mm] \left[\dfrac{1}{\{\sqrt{n}\}}\right], & \text{if } n \text{ is not the square of an integer.} \end{cases}$$

(Here $[x]$ denotes the maximum integer not exceeding x, and $\{x\} = x - [x]$.) Find the value of $\displaystyle\sum_{k=1}^{200} f(k)$.

Solution   For arbitrary $a, k \in \mathbf{N}^+$, if $k^2 < a < (k+1)^2$, we set

$$a = k^2 + m, \ m = 1, \ 2, \ \cdots, \ 2k,$$

$$\sqrt{a} = k + \theta, \ 0 < \theta < 1,$$

then
$$\left[\frac{1}{\{\sqrt{a}\}}\right] = \left[\frac{1}{\sqrt{a} - k}\right] = \left[\frac{\sqrt{a} + k}{a - k^2}\right] = \left[\frac{2k + \theta}{m}\right].$$

For
$$0 < \frac{2k + \theta}{m} - \frac{2k}{\theta} < 1,$$

if there exists an integer t between $\dfrac{2k}{m}$ and $\dfrac{2k + \theta}{m}$, then

$$\frac{2k}{m} < t < \frac{2k + \theta}{m}.$$

On one hand $2k < mt$, thus $2k + 1 \leqslant mt$. On the other hand,

$$mt < 2k + \theta < 2k + 1, \text{ a contradiction.}$$

Thus
$$\left[\frac{2k + \theta}{m}\right] = \left[\frac{2k}{m}\right],$$

$$\sum_{k < a < k+1^2} \left[\frac{1}{\{a\}}\right] = \sum_{m=1}^{2k} \left[\frac{2k}{m}\right],$$

$$\sum_{a=1}^{n+1^2} f(a) = \sum_{k=1}^{n} \sum_{i=1}^{2k} \left[\frac{2k}{i}\right]. \tag{1}$$

Next, we calculate $\displaystyle\sum_{i=1}^{2k} \left[\frac{2k}{i}\right]$. Draw a $2k \times 2k$ table, and put a star $*$ at each place in the i-th row where it is a multiple of i. Then in this row there are $\left[\dfrac{2k}{i}\right]$ stars and the total number of stars in the table is $\displaystyle\sum_{i=1}^{2k} \left[\frac{2k}{i}\right]$. On the other hand, in j-th column, if j has $T(j)$ positive factors, then there are $T(j)$ stars in this column. Thus the total number of stars is $\displaystyle\sum_{j=1}^{2k} T(j)$ and $\displaystyle\sum_{i=1}^{2k} \left[\frac{2k}{i}\right] = \sum_{j=1}^{2k} T(j)$.

For example:

i \ j	1	2	3	4	5	6
1	*	*	*	*	*	*
2		*		*		*
3			*			*
4				*		
5					*	
6						*

Thus

$$\sum_{a=1}^{(n+1)^2} f(a) = \sum_{k=1}^{n} \sum_{j=1}^{2k} T(j)$$

$$= n[T(1)+T(2)]+(n-1)[T(3)+T(4)]$$

$$+\cdots+[T(2n-1)+T(2n)] \tag{2}$$

From (2), $\quad \displaystyle\sum_{k=1}^{16^2} f(k) = \sum_{k=1}^{15}(16-k)[T(2k-1)+T(2k)]. \tag{3}$

Set $a_k = T(2k-1)+T(2k)$, $k=1, 2, \cdots, 15$. It is easy to see that a_k takes the following values.

k	1	2	3	4	5	6	7	8	9	10	11	12	13	14	15
a_k	3	5	6	6	7	8	6	9	8	8	8	10	7	10	10

Therefore, $\quad \displaystyle\sum_{k=1}^{256} f(k) = \sum_{k=1}^{15}(16-k)a_k = 783. \tag{4}$

Note that $f(256) = f(16^2) = 0$ by definition. When $k \in \{241, 242, \cdots, 255\}$, denote $k = 15^2 + r(16 \leqslant r \leqslant 30)$, then

$$\sqrt{k} - 15 = \sqrt{15^2 + r} - 15 = \frac{r}{\sqrt{15^2 + r} + 15},$$

$$\frac{r}{31} < \frac{r}{\sqrt{15^2 + r} + 15} < \frac{r}{30}, \ 1 \leqslant \frac{30}{r} < \frac{1}{\{\sqrt{15^2 + r}\}} < \frac{31}{r} < 2.$$

Thus $$\left[\frac{1}{\{\sqrt{k}\}}\right] = 1, \ k \in \{241, 242, \cdots, 255\}. \qquad (5)$$

Therefore, $$\sum_{k=1}^{200} f(k) = 783 - \sum_{k=201}^{256} f(k) = 783 - 15 = 768.$$

China Mathematical Olympiad

The China Mathematical Olympiad, organized by the China Mathematical Olympiad Committee, is held in January every year. About 150 winners of the China Mathematical Competition take part in it. The competition lasts for 2 days, and there are 3 problems to be completed within 4.5 hours each day.

2003 (Changsha, Hunan)

2003 China Mathematical Olympiad and 18th Mathematics Winter Camp was held on January 11 – 18 in Changsha, Hunan Province, and was hosted by Hunan Mathematical Committee and Changsha No. 1 middle school.

The Competition Committee consisted of the following: Xu Yichao, Li Shenghong, Shen Wenxuan, Chen Yonggao, Su Chun, Leng Gangsong, Huang Yumin, Huang Xuanguo.

First Day
8:00 – 12:30 January 15, 2003

① Suppose points I and H are the incenter and orthocenter of an acute triangle ABC respectively, and points B_1 and C_1 are the midpoints of sides AC and AB respectively. It is known that ray $B_1 I$ intersects side AB at $B_2 (B_2 \neq B)$, and ray $C_1 I$ intersects the extension of AC at C_2: $B_2 C_2$ and BC intersect at K and A_1 is the circumcenter of $\triangle BHC$. Prove that three points A, I and A_1 are collinear if and only if the areas of $\triangle BKB_2$ and $\triangle CKC_2$ are equal. (posed by Shen Wenxuan)

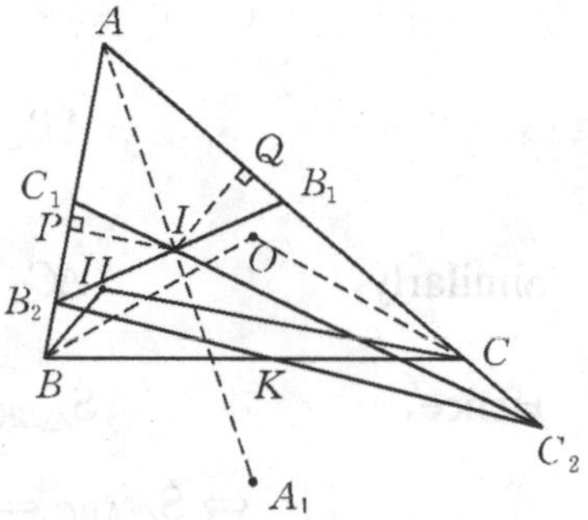

Proof First, we will prove that three points A, I and A_1 are collinear $\Leftrightarrow \angle BAC = 60°$.

As shown in the figure, assume that O is the circumcenter of $\triangle ABC$. We join BO and CO, then

$$\angle BHC = 180° - \angle BAC,$$

$$\angle BA_1 C = 2(180° - \angle BHC) = 2\angle BAC.$$

Hence, $\angle BAC = 60° \Leftrightarrow \angle BAC + \angle BA_1 C = 180°$

$\Leftrightarrow A_1$ is on the circumcircle $\odot O$ of $\triangle ABC$

$\Leftrightarrow AI$ and AA_1 coincide (because A_1 is on the perpendicular bisector of BC.)

$\Leftrightarrow$ Three points A, I and A_1 are collinear.

Secondly, we will prove that $S_{\triangle BKB_2} = S_{\triangle CKC_2} \Leftrightarrow \angle BAC = 60°$.

Construct $IP \perp AB$ at point P, and $IQ \perp AC$ at Q, then

$$S_{\triangle AB_1B_2} = \frac{1}{2}IP \cdot AB_2 + \frac{1}{2}IQ \cdot AB_1.$$

Note that $\quad S_{\triangle AB_1B_2} = \frac{1}{2}AB_1 \cdot AB_2 \cdot \sin A,$

therefore $\quad IP \cdot AB_2 + IQ \cdot AB_1 = AB_1 \cdot AB_2 \cdot \sin A.$

Assume $IP = r$, where r is the radius of the inscribed circle of $\triangle ABC$. Then $IQ = r$. Furthermore set $BC = a$, $CA = b$ and $AB = c$, then $r = \dfrac{2S_{\triangle ABC}}{a+b+c}$.

From $AB_1 = \dfrac{b}{2}$ and $\quad 2AB_1 \cdot \sin A = h_c = \dfrac{2S_{\triangle ABC}}{c}$,

we have $\quad AB_2 \cdot \left(\dfrac{2S_{\triangle ABC}}{c} - 2 \cdot \dfrac{2S_{\triangle ABC}}{a+b+c} \right) = b \cdot \dfrac{2S_{\triangle ABC}}{a+b+c}$,

so $\qquad\qquad\qquad AB_2 = \dfrac{bc}{a+b+c}.$

Similarly, $\qquad\qquad AC_2 = \dfrac{bc}{a+c-b}.$

Hence, $\qquad\qquad\quad S_{\triangle BKB_2} = S_{\triangle CKC_2}$

$$\Leftrightarrow S_{\triangle ABC} = S_{\triangle AB_2C_2}$$

$$\Leftrightarrow bc = \dfrac{bc}{a+b-c} \cdot \dfrac{bc}{a+c-b}$$

$$\Leftrightarrow a^2 = b^2 + c^2 - bc$$

$$\Leftrightarrow \angle BAC = 60° \text{ (By the law of cosines)}.$$

Therefore, the proposition is true.

② Suppose a set S satisfies the following conditions:
   (1) every element in S is a positive integer and not greater than 100;

(2) for any two different elements a and b in S, there is an element c in S such that the greatest common divisor of a and c is equal to 1, and the greatest common divisor of b and c is also 1; and

(3) for any two different elements a and b in S, there is an element d, which is different from a and b, such that the greatest common divisor of a and d, and that of b and d are greater than 1.

Find the maximum number of elements in S. (posed by Yao Jiangang)

Solution   The maximum number of elements is 72.

A positive integer not greater than 100 can be written as

$$n = 2^{\alpha_1} \cdot 3^{\alpha_2} \cdot 5^{\alpha_3} \cdot 7^{\alpha_4} \cdot 11^{\alpha_5} \cdot q,$$

where q is a positive integer and not divisible by 2, 3, 5, 7 and 11, and α_1, α_2, α_3, α_4 and α_5 are nonnegative integers.

We pick out those positive integers n with just one or two nonzero among α_1, α_2, α_3, α_4 and α_5 to form set S. In this case, S contains 50 even numbers (2, 4, $\cdots$, 98 and 100) except the following seven: $2 \times 3 \times 5$, $2^2 \times 3 \times 5$, $2 \times 3^2 \times 5$, $2 \times 3 \times 7$, $2^2 \times 3 \times 7$, $2 \times 5 \times 7$ and $2 \times 3 \times 11$, 17 odd numbers that are multiples of three (i. e. 3×1, 3×3, $\cdots$, 3×33), 7 odd numbers with the least prime divisor 5 (i. e. 5×1, 5×5, 5×7, 5×11, 5×13, 5×17 and 5×19), 4 odd numbers with the least prime divisor 7 (i. e. 7×1, 7×7, 7×11 and 7×13), and the prime number 11.

Consequently, S contains $(50 - 7) + 17 + 7 + 4 + 1 = 72$ numbers totally.

In what follows, we will prove that S constructed above satisfies the given condition.

Obviously, it satisfies condition (1).

For condition (2), we note that, at most, four prime divisors among 2, 3, 5, 7 and 11 will occur in $[a, b]$. We write the prime which does not occur, as p. Obviously, $p \in S$ and

$$(p, a) \leqslant (p, [a, b]) = 1,$$

$$(p, b) \leqslant (p, [a, b]) = 1.$$

Hence, we take $c = p$.

For condition (3), we take the least prime divisor of a as p and the one of b as q when $(a, b) = 1$. It is easy to see that $p \neq q$ and p, $q \in \{2, 3, 4, 7, 11\}$. Hence $pq \in S$, and

$$(pq, a) \geqslant p > 1 \text{ and } (pq, b) \geqslant q > 1.$$

Being coprime to each other for a and b ensures that pq is different from a and b. Thus we take $d = pq$.

When $(a, b) = e > 1$, we take p as the least prim divisor of e, and q as the smallest prime number satisfying $q \nmid [a, b]$. It is easy to see that $p \neq q$, and p and $q \in \{2, 3, 5, 7, 11\}$. Hence $pq \in S$, and

$$(pq, a) \geqslant (p, a) = p > 1,$$

$$(pq, b) \geqslant (p, b) = p > 1.$$

$q \nmid [a, b]$ ensures that pq is different from a and b. Thus, we take $d = pq$.

In what follows, we prove that the number of elements in S, which satisfies the conditions described in the problem, will not be greater than 72.

Obviously, $1 \notin S$. For arbitrary two prime numbers p and q which are both greater than 10, since the least number which is not prime to neither p nor q is pq, it must be greater than 100. So we know, according to condition (3), that there is at most one among 21 prime numbers between 10 and 100 (11, 13, $\cdots$, 89, 97), occurring in S. We write the set consisting of all natural numbers not greater than 100 except 1 and the above-mentioned 21 prime numbers as T, and there are 78 numbers in the set. We can conclude that there are at least 7 numbers in T are not in S. Thus S contains at most $78 - 7 + 1 = 72$ elements.

(i) When a prime number p is greater than 10 and belongs to S, every number in S can only have 2, 3, 5, 7 and p as its least prime divisor. By condition (2), we have the following conclusions.

① If $7p \in S$, because $\{2 \times 3 \times 5, 2^2 \times 3 \times 5, 2 \times 3^2 \times 5, 7p\}$ contains all the least prime divisors, we know from condition (2) that $2 \times 3 \times 5$, $2^2 \times 3 \times 5$ and $2 \times 3^2 \times 5$ do not belong to S. If $7p \notin S$, noting $2 \times 7p > 100$, but $p \in S$, so from condition (2) we know that 7×1, 7×7, 7×11 and 7×13 do not belong to S.

② If $5p \in S$, then $2 \times 3 \times 7$ and $2^2 \times 3 \times 7$ do not belong to S. If $5p \notin S$, then 5×1 and 5×5 do not belong to S.

③ $2 \times 5 \times 7$ and $3p$ do not belong to S at the same time.

④ $2 \times 3p$ and 5×7 do not belong to S at the same time.

⑤ If $5p$, $7p \notin S$, then $5 \times 7 \notin S$.

When $p = 11$ or 13, from ①, ②, ③ and ④, we can get at least 3, 2, 1 and 1 numbers in T respectively which do not belong to S, and in total 7 numbers. When $p = 17$ or 19, from ①, ② and ③, we can get at least 4, 2 and 1 numbers in T respectively, which do not belong to S and in total 7 numbers. When $p > 20$, from ①, ② and ③, there are at least 4, 2 and 1 numbers in T respectively, which do not belong to S and in total 7 numbers also.

(ii) If there is no prime number greater than 10 belonging to S, then the least prime numbers in S can only be 2, 3, 5 and 7. Hence, each of the following 7 pairs of numbers can not belong to S at the same time:

$$(3, 2 \times 5 \times 7), (5, 2 \times 3 \times 7), (7, 2 \times 3 \times 5), (2 \times 3, 5 \times 7),$$

$$(2 \times 5, 3 \times 7), (2 \times 7, 3 \times 5), (2^2 \times 7, 3^2 \times 5).$$

Thus, there are at least 7 numbers in T that are not in S.

Consequently, the answer for this problem is 72.

● Given a positive integer n, find the least positive number λ such that $\cos \theta_1 + \cos \theta_2 + \cdots + \cos \theta_n$ is not greater than λ provided $\tan \theta_1 \cdot \tan \theta_2 \cdot \cdots \cdot \tan \theta_n = 2^{\frac{n}{2}}$ for any $\theta_i \in \left(0, \dfrac{\pi}{2}\right)$ $(i = 1, 2, \cdots,$

n). (posed by Huang Yumin)

Solution   When $n = 1$, $\cos \theta_1 = (1 + \tan^2 \theta_1)^{-\frac{1}{2}} = \dfrac{\sqrt{3}}{3}$. Hence $\lambda = \dfrac{\sqrt{3}}{3}$.

When $n = 2$, we can prove

$$\cos \theta_1 + \cos \theta_2 \leqslant \frac{2\sqrt{3}}{3},$$ ①

and when $\theta_1 = \theta_2 = \arctan \sqrt{2}$, the equality holds.

In fact,

$$① \Leftrightarrow \cos^2 \theta_1 + \cos^2 \theta_2 + 2\cos \theta_1 \cdot \cos \theta_2 \leqslant \frac{3}{4},$$

that is, $\dfrac{1}{1 + \tan^2 \theta_1} + \dfrac{1}{1 + \tan^2 \theta_2} + 2 \sqrt{\dfrac{1}{(1 + \tan^2 \theta_1)(1 + \tan^2 \theta_2)}} \leqslant \dfrac{3}{4}.$

②

From $\tan \theta_1 \cdot \tan \theta_2 = 2$, we get

$$② \Leftrightarrow \frac{2 + \tan^2 \theta_1 + \tan^2 \theta_2}{5 + \tan^2 \theta_1 + \tan^2 \theta_2} + 2 \sqrt{\frac{1}{5 + \tan^2 \theta_1 + \tan^2 \theta_2}} \leqslant \frac{3}{4}.$$ ③

Write $x = \tan^2 \theta_1 + \tan^2 \theta_2$, then

$$③ \Leftrightarrow 2 \sqrt{\frac{1}{5 + x}} \leqslant \frac{14 + x}{3(5 + x)},$$

that is, $36(5 + x) \leqslant 196 + 28x + x^2.$ ④

Obviously, $④ \Leftrightarrow x - 8x + 16 = (x - 4)^2 \geqslant 0.$

Hence, $\lambda = \dfrac{2\sqrt{3}}{3}.$

When $n \geqslant 3$, there is no loss of generality in supposing $\theta_1 \geqslant \theta_2 \geqslant \cdots \geqslant \theta_n$, then

$$\tan \theta_1 \cdot \tan \theta_2 \cdot \tan \theta_3 \geqslant 2\sqrt{2}.$$

Since $\cos \theta_i = \sqrt{1 - \sin^2 \theta_i} < 1 - \dfrac{1}{2}\sin^2 \theta_i$, so

$$\cos \theta_2 + \cos \theta_3 < 2 - \frac{1}{2}(\sin^2 \theta_2 + \sin^2 \theta_3)$$

$$< 2 - \sin \theta_2 \cdot \sin \theta_3.$$

From $\tan^2 \theta_1 \geqslant \dfrac{8}{\tan^2 \theta_2 \cdot \tan^2 \theta_3}$, we have

$$\frac{1}{\cos^2 \theta_1} \geqslant \frac{8 + \tan^2 \theta_2 \cdot \tan^2 \theta_3}{\tan^2 \theta_2 \cdot \tan^2 \theta_3},$$

that is,
$$\cos \theta_1 \leqslant \frac{\tan \theta_2 \cdot \tan \theta_3}{\sqrt{8 + \tan^2 \theta_2 \cdot \tan^2 \theta_3}}$$

$$= \frac{\sin \theta_2 \cdot \sin \theta_3}{\sqrt{8\cos^2 \theta_2 \cdot \cos^2 \theta_3 + \sin^2 \theta_2 \cdot \sin^2 \theta_3}}.$$

Hence
$$\cos \theta_2 + \cos \theta_3 + \cos \theta_1$$

$$< 2 - \sin \theta_2 \cdot \sin \theta_3 \left[1 - \frac{1}{\sqrt{8\cos^2 \theta_2 \cdot \cos^2 \theta_3 + \sin^2 \theta_2 \cdot \sin^2 \theta_3}} \right].$$

⑤

Note that
$$8\cos^2 \theta_2 \cdot \cos^2 \theta_3 + \sin^2 \theta_2 \cdot \sin^2 \theta_3 \geqslant 1$$

$$\Leftrightarrow 8 + \tan^2 \theta_2 \tan^2 \theta_3 \geqslant \frac{1}{\cos^2 \theta_2 \cdot \cos^2 \theta_3}$$

$$= (1 + \tan^2 \theta_2)(1 + \tan^2 \theta_3)$$

$$\Leftrightarrow \tan^2 \theta_2 + \tan^2 \theta_3 \leqslant 7.$$

⑥

Thus, we get that, when ⑥ **holds,**

$$\cos \theta_1 + \cos \theta_2 + \cos \theta_3 < 2.$$

⑦

If ⑥ **does not hold, then** $\tan^2 \theta_2 + \tan^2 \theta_3 > 7,$

hence $$\tan^2 \theta_1 \geqslant \tan^2 \theta_2 > \frac{7}{2},$$

so $$\cos \theta_1 \leqslant \cos \theta_2 < \sqrt{\frac{1}{1 + \frac{7}{2}}} = \frac{\sqrt{2}}{3}.$$

Thus $$\cos \theta_1 + \cos \theta_2 + \cos \theta_3 < \frac{2\sqrt{2}}{3} + 1 < 2,$$

that is, ⑦ holds too.

Hence we can get

$$\cos \theta_1 + \cos \theta_2 + \cos \theta_3 + \cdots + \cos \theta_n < n - 1.$$

On the other hand, if we take $\theta_2 = \theta_3 = \cdots = \theta_n = \alpha > 0$, $\alpha \to 0$, then

$$\theta_1 = \arctan \frac{2^{\frac{n}{2}}}{(\tan \alpha)^{n-1}}.$$

Obviously, $\theta_1 \to \frac{\pi}{2}$, thus

$$\cos \theta_1 + \cos \theta_2 + \cos \theta_3 + \cdots + \cos \theta_n \to n - 1.$$

Consequently, we get $\lambda = n - 1$.

Second Day
8:00 – 12:30 January 16, 2003

⬤ Find all ternary positive integer groups (a, m, n) satisfying $a \geqslant 2$ and $m \geqslant 2$ such that $a^n + 203$ is a multiple of $a^m + 1$. (posed by Chen Yonggao)

Solution We will discuss the following three cases for n and m.

(i) In the case when $n < m$, from $a^n + 203 \geqslant a^m + 1$, we have

$$202 \geqslant a^m - a^n \geqslant a^n(a - 1) \geqslant a(a - 1).$$

Therefore, $$2 \leqslant a \leqslant 14.$$

When $a = 2$, we can take n to be 1, 2, $\cdots$, 7.

When $a = 3$, we can take n to be 1, 2, 3 and 4.

When $a = 4$, we can take n to be 1, 2 and 3.

When $5 \leqslant a \leqslant 6$, we can take n to be 1 and 2.

When $7 \leqslant a \leqslant 14$, $n = 1$.

From $a^m + 1 \mid a^n + 203$, we can deduce that the solutions are $(2, 2, 1)$, $(2, 3, 2)$ and $(5, 2, 1)$.

(ii) In the case when $n = m$, $a^m + 1 \mid 202$. Since 202 has only four divisors 1, 2, 101 and 202, but $a \geqslant 2$, $m \geqslant 2$ and $a^m + 1 \geqslant 5$, so $a^m = 100$ or 201. Since $m \geqslant 2$, therefore the solution is $(10, 2, 2)$.

(iii) In the case when $n > m$, from $a^m + 1 \mid 203(a^m + 1)$, we have

$$a^m + 1 \mid a^n + 203 - (203a^m + 203),$$

that is, $$a^m + 1 \mid a^m(a^{n-m} - 203).$$

Since $(a^m + 1, a^m) = 1$, so

$$a^m + 1 \mid a^{n-m} - 203.$$

① If $a^{n-m} < 203$, then set $n - m = s \geqslant 1$, we have $a^m + 1 \mid 203 - a^s$. Hence

$$203 - a^s \geqslant a^m + 1,$$

$$202 \geqslant a^s + a^m \geqslant a^m + a$$

$$= a(a^{m-1} + 1) \geqslant a(a + 1),$$

thus $$2 \leqslant a \leqslant 13.$$

Using the same orgument as in Case (i), we show that the solutions for (a, m, s) are:

$$(2, 2, 3), (2, 6, 3), (2, 4, 4), (2, 3, 5), (2, 2, 7),$$

$$(3, 2, 1), (4, 2, 2), (5, 2, 3) \text{ and } (8, 2, 1).$$

Hence, (a, m, n) are

$$(2, 2, 5), (2, 6, 9), (2, 4, 8), (2, 3, 8), (2, 2, 9),$$

$$(3, 2, 3), (4, 2, 4), (5, 2, 5) \text{ and } (8, 2, 3).$$

② If $a^{n-m} = 203$, then $a = 203$ and $n-m = 1$, and the solution is

$$(203, m, m+1), m \geqslant 2.$$

③ If $a^{n-m} > 203$, set $n-m = s \geqslant 1$, then $a^m + 1 \mid a^s - 203$.
Since $a^s - 203 \geqslant a^m + 1$, so $s > m$. By

$$a^m + 1 \mid a^s + 203a^m = (a^{s-m} + 203)a^m$$
$$= (a^{n-2m} + 203)a^m,$$

and $(a^m + 1, a^m) = 1,$

we have $a^m + 1 \mid a^{n-2m} + 203.$

Now $s > m \Leftrightarrow n - m > m \Leftrightarrow n > 2m \Leftrightarrow n - 2m > 0$. In this case, the solutions can only be derived from the preceding solutions, that is, from $(a, m, n) \to (a, m, n+2m) \to \cdots \to (a, m, n+2km)$, and each solution derived also satisfies $a^m + 1 \mid a^n + 203$.

Summarizing what described above, we obtair all solutions (a, m, n) to be:

$$(2, 2, 4k+1), (2, 3, 6k+2), (2, 4, 8k+8), (2, 6, 12k+9),$$

$$(3, 2, 4k+3), (4, 2, 4k+4), (5, 2, 4k+1), (8, 2, 4k+3),$$

$$(10, 2, 4k+2) \text{ and } (203, m, (2k+1)m+1),$$

where k is any nonnegative integer, and $m \geqslant 2$ is an integer.

🌑 A certain company wants to employ one secretary. Ten personsapply. The manager decides to interview them one by one according to the order of their applications. The first 3 applicants should not be employed. From the fourth onward an applicant will be compared with the preceding ones. If he exceeds in ability all the procedings applicants, he will be employed. Otherwise he will not, and the interview goes on. If the preceding nine persons are not employed, the last one will be employed.

Suppose that the 10 persons are different from each other in ability, and we can arrange them according to their ability rating from superior to inferior, such as 1st, 2nd, ⋯, 10th. Obviously, whether an applicant will be eventually employed by the company depends on the order of the applications. As it is known, there are 10! such permutations in all. Now denote by A_k the number of the permutations such that the applicant with the k-th ability rating is employed and the probability for him to be employed is $\dfrac{A_k}{10!}$. (posed by Su Chun)

Prove that under the policy given by the manager, we have the following properties:

(1) $A_1 > A_2 > \cdots > A_8 = A_9 = A_{10}$.

(2) The probability for the company to employ one of the persons with the ability among the three is over 70%, and to employ one of the persons with the ability among the bottom three is not over 10%. (provided by Su Chun)

Proof We denote by a the ability rating of the applicant with the highest ability among the first three interviews. Obviously, $a \leqslant 8$. Now we denote by $A_k(a)$ the set of permutations for which the person with the k-th ability rating is employed and we denote the corresponding number of permutations by $|A_k(a)|$.

(1) It is easy to see that, when $a = 1$, a definite result is to give up the first nine persons and to employ the last interviewee. Now the chance is equal for every person except the one who has the highest ability rating. It is not difficult to see that

$$|A_k(1)| = 3 \times 8! := r_1, \quad k = 2, 3, \cdots, 10,$$

where ":=" denotes "writter as".

When $2 \leqslant a \leqslant 8$, a person who is placed k-th in ability rating will have no chance to be employed for $a \leqslant k \leqslant 10$, and the chance is equal for $1 \leqslant k < a$.

In fact, a person with the a-th ability rating is among the first three interviews and there are three ways to choose one out of the three interviews. Those people with their ability rating from the 1-st to $(a-1)$-th are among the later seven interviews, and the first one having the highest ability rating among them is employed. There are $C_7^{a-1}(a-2)!$ ways of such arrangement. The other $10-a$ persons may be arranged arbitrarily on the remainder positions, and we have $(10-a)!$ ways of arrangement. Hence, we have

$$| A_k(a) | = \begin{cases} 3C_7^{a-1}(a-2)!(10-a)! := r_a, & k = 1, \cdots, a-1; \\ 0, & k = a, \cdots, 10. \end{cases}$$

The result above shows:

$$A_8 = A_9 = A_{10} = r_1 = 3 \times 8! > 0; \qquad \qquad ①$$

$$A_k = r_1 + \sum_{a=k+1}^{8} r_a, \ k = 2, \cdots, 7; \qquad \qquad ②$$

and
$$A_1 = \sum_{a=2}^{8} r_a. \qquad \qquad ③$$

From ① and ②, we obtain

$$A_2 > A_3 > \cdots > A_8 = A_9 = A_{10} > 0;$$

and from ② and ③, we obtain

$$A_1 - A_2 = r_2 - r_1 = 3 \times 7 \times 8! - 3 \times 8! > 0.$$

Consequently, (1) is proved.

(2) From ①, we know

$$\frac{A_8 + A_9 + A_{10}}{10!} = \frac{3 \times r_1}{10!} = \frac{3 \times 3 \times 8!}{10!} = 10\%.$$

So the probability for the company to employ one of the bottom three is equal to 10%.

From ② and ③, we can show that

$$A_1 = \sum_{a=2}^{8} r_a = \sum_{a=2}^{8} 3C_7^{a-1}(a-2)!(10-a)!$$

$$= 3 \times 7! \sum_{a=2}^{8} \frac{(9-a)(10-a)}{a-1}$$

$$=3\times7!\sum_{s=1}^{7}\frac{(8-s)(9-s)}{s}$$

$$=3\times7!\times\left(56+21+10+5+\frac{12}{5}+1+\frac{2}{7}\right)$$

$$=3\times7!\times95\frac{24}{35}>3\times7!\times95\frac{2}{3}$$

$$=287\times7!;$$

$$A_2=r_1+\sum_{a=3}^{8}r_a$$

$$=3\times8!+3\times7!\times\left(21+10+5+\frac{12}{5}+1+\frac{2}{7}\right)$$

$$=3\times7!\times47\frac{24}{35}>3\times7!\times47\frac{2}{3}=143\times7!;$$

$$A_3=r_1+\sum_{a=4}^{8}r_a$$

$$=3\times8!+3\times7!\times\left(10+5+\frac{12}{5}+1+\frac{2}{7}\right)$$

$$=3\times7!\times26\frac{24}{35}>3\times7!\times26\frac{2}{3}=80\times7!.$$

Therefore,

$$\frac{A_1+A_2+A_3}{10!}>\frac{287+143+80}{720}$$

$$=\frac{510}{720}=\frac{17}{24}>70\%.$$

That is, the probability for the company to employ one of the top three, is greater than 70%.

⬤ Suppose a, b, c and d are positive real numbers satisfying $ab+cd=1$ and $P_i(x_i,y_i)$ $(i=1,2,3,4)$ are four points on the unit

circle which has the origin as its center. Prove that:
$$(ay_1 + by_2 + cy_3 + dy_4)^2 + (ax_4 + bx_3 + cx_2 + dx_1)^2 \leqslant$$
$$2\left(\frac{a^2+b^2}{ab} + \frac{c^2+d^2}{cd}\right). \text{ (posed by Li Shenghong)}$$

Proof I Set $u = ay_1 + by_2$, $v = cy_3 + dy_4$, $u_1 = ax_4 + bx_3$ and $v_1 = cx_2 + dx_1$. Then

$$u^2 \leqslant (ay_1 + by_2)^2 + (ax_1 - bx_2)^2$$
$$= a^2 + b^2 + 2ab(y_1 y_2 - x_1 x_2),$$

that is
$$x_1 x_2 - y_1 y_2 \leqslant \frac{a^2+b^2-u^2}{2ab}. \qquad \text{①}$$

$$v_1^2 \leqslant (cx_2 + dx_1)^2 + (cy_2 - dy_1)^2$$
$$= c^2 + d^2 + 2cd(x_1 x_2 - y_1 y_2),$$

that is

$$y_1 y_2 - x_1 x_2 \leqslant \frac{c^2+d^2-v_1^2}{2cd}. \qquad \text{②}$$

①＋②, we get

$$0 \leqslant \frac{a^2+b^2-u^2}{2ab} + \frac{c^2+d^2-v_1^2}{2cd},$$

that is
$$\frac{u^2}{ab} + \frac{v_1^2}{cd} \leqslant \frac{a^2+b^2}{ab} + \frac{c^2+d^2}{cd}.$$

Similarly
$$\frac{v^2}{cd} + \frac{u_1^2}{ab} \leqslant \frac{c^2+d^2}{cd} + \frac{a^2+b^2}{ab}.$$

By Cauchy's inequality, we have
$$(u+v)^2 + (u_1+v_1)^2$$
$$\leqslant (ab+cd)\left(\frac{u^2}{ab} + \frac{v^2}{cd}\right) + (ab+cd)\left(\frac{u_1^2}{ab} + \frac{v_1^2}{cd}\right)$$
$$= \frac{u^2}{ab} + \frac{v^2}{cd} + \frac{u_1^2}{ab} + \frac{v_1^2}{cd} \leqslant 2\left[\frac{a^2+b^2}{ab} + \frac{c^2+d^2}{cd}\right].$$

Proof II   By Cauchy's inequality, we can show

$$(ay_1 + by_2 + cy_3 + dy_4)^2$$

$$\leqslant (ab + cd) \left[\frac{(ay_1 + by_2)^2}{ab} + \frac{(cy_3 + dy_4)^2}{cd} \right]$$

$$= \frac{a}{b} y_1^2 + \frac{b}{a} y_2^2 + \frac{c}{d} y_3^2 + \frac{d}{c} y_4^2 + 2(y_1 y_2 + y_3 y_4).$$

Similarly, $(ax_4 + bx_3 + cx_2 + dx_1)^2$

$$\leqslant \frac{a}{b} x_4^2 + \frac{b}{a} x_3^2 + \frac{c}{d} x_2^2 + \frac{d}{c} x_1^2 + 2(x_1 x_2 + x_3 x_4).$$

So we subtract the right-hand side (RHS) from the left-hand side (LHS) in the original inequality and get

$$\text{LHS} - \text{RHS} \leqslant \frac{a}{b} y_1^2 + \frac{b}{a} y_2^2 + \frac{c}{d} y_3^2 + \frac{d}{c} y_4^2 + 2y_1 y_2 + 2y_3 y_4 +$$

$$\frac{a}{b} x_4^2 + \frac{b}{a} x_3^2 + \frac{c}{d} x_2^2 + \frac{d}{c} x_1^2 + 2x_1 x_2 + 2x_3 x_4 -$$

$$2 \left(\frac{a}{b} + \frac{b}{a} + \frac{c}{d} + \frac{d}{c} \right).$$

$$= -\frac{a}{b} x_1^2 - \frac{b}{a} x_1^2 - \frac{c}{d} x_3^2 - \frac{d}{c} x_4^2 - \frac{a}{b} y_4^2 - \frac{b}{a} y_3^2 -$$

$$\frac{c}{d} y_2^2 - \frac{d}{c} y_1^2 + 2(x_1 x_2 + x_3 x_4 + y_1 y_2 + y_3 y_4)$$

$$\leqslant -2x_1 x_2 - 2x_3 x_4 - 2y_1 y_2 - 2y_3 y_4 +$$

$$2(x_1 x_2 + x_3 x_4 + y_1 y_2 + y_3 y_4)$$

$$= 0.$$

The proposition is proved.

2004 (Macao)

2004 ChinaMathematical Olympiad and 19th Mathematics Winter Camp(for secondary school students) was held on January 6 – 11 in Macao, and was hosted by Education and Youth Affair Bureau of Macao Special Administrative Region of PRC.

The Competition Committee consisted of the following: Chen Yonggao, Liang Yingde, Huang Yumin, Li Shenghong, Yu Hongbing, Li Weigu, Qin Hourong, Xiong Bin, Wang Jianwei, and Xu Jiangxiong.

First Day
8:30 – 13:00   January 8, 2004

⬤ Let $EFGH$, $ADCD$ and $E_1 F_1 G_1 H_1$ be three convex quadrilaterals, satisfying: (a) Points E, F, G and H lie on sides AB, BC, CD and DA, respectively, and $\dfrac{AE}{EB} \cdot \dfrac{BF}{FC} \cdot \dfrac{CG}{GD} \cdot \dfrac{DH}{HA} = 1$; (b) points A, B, C and D lie on sides $H_1 E_1$, $E_1 F_1$, $F_1 G_1$ and $G_1 H_1$, respectively, and $E_1 F_1 \parallel EF$, $F_1 G_1 \parallel FG$, $G_1 H_1 \parallel GH$, $H_1 E_1 \parallel HE$. Suppose $\dfrac{E_1 A}{AH_1} = \lambda$, find the expression of $\dfrac{F_1 C}{CG_1}$ in terms of λ.

(posed by Xiong Bin)

Solution (1) If $EF \parallel AC$, then $\dfrac{BE}{EA} = \dfrac{BF}{FC}$.

So $\dfrac{DH}{HA} = \dfrac{DG}{GC}$ using condition (a). Then

$HG \parallel AC$, giving $E_1 F_1 \parallel AC \parallel H_1 G_1$. That means $\dfrac{F_1C}{CG_1} = \dfrac{E_1A}{AH_1} = \lambda.$

(2) If EF is not parallel to AC, extend lines EF and AC to meet at point T. By Menelaus' Theorem, we have $\dfrac{CF}{FB} \cdot \dfrac{BE}{EA} \cdot$ $\dfrac{AT}{TC} = 1$, and then $\dfrac{CG}{GD} \cdot \dfrac{DH}{HA} \cdot \dfrac{AT}{TC} = 1$ using condition (a). By the inverse of Menelaus' Theorem, we know that points T, H and G are collinear. Suppose lines TF and TG meet line E_1H_1 at M and N respectively. As $EB_1 \parallel EF$, we get $E_1A = \dfrac{BA}{EA} \cdot AM$. In the same way, we get $H_1A = \dfrac{AD}{AH} \cdot AN$. Then

$$\frac{E_1A}{H_1A} = \frac{AM}{AN} \cdot \frac{AB}{AE} \cdot \frac{AH}{AD}. \qquad ①$$

On the other hand, $\dfrac{AM}{AN} = \dfrac{EQ}{QH} = \dfrac{\triangle AEC}{\triangle AHC} = \dfrac{\triangle ABC}{\triangle ADC} \cdot \dfrac{AE}{AB} \cdot \dfrac{AD}{AH}.$

$$② $$

From ①, ② we get $\dfrac{E_1A}{H_1A} = \dfrac{EQ}{QH} \cdot \dfrac{AB}{AE} \cdot \dfrac{AH}{AD} = \dfrac{\triangle ABC}{\triangle ADC}.$ In the same way,

$$\frac{F_1C}{CG_1} = \frac{\triangle ABC}{\triangle ADC}.$$

So $\dfrac{F_1C}{CG_1} = \dfrac{E_1A}{AH_1} = \lambda.$

Let c be a positive integer, and a number sequence x_1, x_2, $\cdots$ satisfy $x_1 = c$ and

$$x_n = x_{n-1} + \left[\frac{2x_{n-1} - (n+2)}{n} \right] + 1, \; n = 2, 3, \cdots,$$

where $[x]$ denotes the largest integer not greater than x.

Determine the expression of x_n in terms of n and c. (posed by Huang Yumin)

Solution   Obviously, $x_n = x_{n-1} + \left[\dfrac{2(x_{n-1}-1)}{n}\right]$ for $n \geqslant 2$. Let $a_n = x_n - 1$, then $a_1 = c - 1$,

$$a_n = a_{n-1} + \left[\frac{2a_{n-1}}{n}\right] = \left[\frac{n+2}{n}a_{n-1}\right], \ n = 2, 3, \cdots. \qquad ①$$

Let $u_n = A\dfrac{(n+1)(n+2)}{2}$, $n = 1, 2, \cdots$, where A is a nonnegative integer. Since

$$\left[\frac{n+2}{n}u_{n-1}\right] = \left[A\frac{n+2}{2n}n(n+1)\right]$$

$$= A\frac{(n+1)(n+2)}{2} = u_n, \text{ for } n \geqslant 2,$$

the sequence $\{u_n\}$ satisfies ①.

Let $y_n = n$, $n = 1, 2, \cdots$. Since

$$\left[\frac{n+2}{n}y_{n-1}\right] = \left[\frac{(n+2)(n-1)}{n}\right]$$

$$= \left[n+1-\frac{2}{n}\right] = y_n, \text{ for } n \geqslant 2,$$

the sequence $\{y_n\}$ satisfies ① too.

Let $z_n = \left[\dfrac{(n+2)^2}{4}\right]$, $n = 1, 2, \cdots$. Then we have, for $n = 2m$ and $m \geqslant 1$,

$$\left[\frac{n+2}{n}z_{n-1}\right] = \left[\frac{m+1}{m}\left[\frac{(2m+1)^2}{4}\right]\right]$$

$$= \left[\frac{m+1}{m}m(m+1)\right] = (m+1)^2 = z_n.$$

For $n = 2m+1$ and $m \geqslant 1$,

$$\left[\frac{n+2}{n}z_{n-1}\right] = \left[\frac{2m+3}{2m+1}\left[\frac{(2m+2)^2}{4}\right]\right] = \left[\frac{2m+3}{2m+1}(m+1)^2\right]$$

$$= \left[(m+1)(m+2)+\frac{m+1}{2m+1}\right] = (m+1)(m+2)$$

$$= \left[\frac{(2m+3)^2}{4}\right] = z_n.$$

So, the sequence $\{y_n\}$ satisfies ① too.

For any nonnegative integer A, let $v_n = u_n + y_n = A \cdot \frac{(n+1)(n+2)}{2} + n$,

$$w_n = u_n + z_n = A \cdot \frac{(n+1)(n+2)}{2} + \left[\frac{(n+2)^2}{4}\right],$$

$$n = 1, 2, \cdots.$$

Obviously, both $\{v_n\}$ and $\{w_n\}$ satisfy ①.

Since $u_1 = 3A$, $y_1 = 1$, $z_1 = \left[\frac{9}{4}\right] = 2$, then for $3 \mid a_1$ we have

$$a_n = \frac{a_1}{6}(n+1)(n+2).$$

For $a_1 \equiv 1 \pmod 3$, $a_n = \frac{a_1-1}{6}(n+1)(n+2)+n$. For $a_1 \equiv 2 \pmod 3$,

$$a_n = \frac{a_1-2}{6}(n+1)(n+2)+\left[\frac{(n+2)^2}{4}\right].$$

In summary,

$$x_n = \frac{c-1}{6}(n+1)(n+2)+1, \text{ for } c \equiv 1 \pmod 3;$$

$$x_n = \frac{c-2}{6}(n+1)(n+2)+n+1, \text{ for } c \equiv 2 \pmod 3;$$

$$x_n = \frac{c-3}{6}(n+1)(n+2)+\left[\frac{(n+2)^2}{4}\right]+1, \text{ for } c \equiv 0 \pmod 3.$$

⚙ Let M be a set consisting of n points in the plane, and satisfying:
(1) there exist 7 points in M which constitute the vertices of a convex heptagon; (2) if for any 5 points in M which constitute the vertices of a convex pentagon, then there is a point in M which lies in the interior of the pentagon.

Find the minimum value of n. (posed by Leng Gangsong)

Solution First, we prove that $n \geqslant 11$. Suppose a convex heptagon has its vertices in M given by $A_1 A_2 A_3 A_4 A_5 A_6 A_7$. Using Condition (1), we get that there exists one point P_1 belonging to M in the interior of convex pentagon $A_1 A_2 A_3 A_4 A_5$. Connecting $P_1 A_1$ and $P_1 A_5$, we obtain that there exists one point P_2 in convex pentagon $A_1 P_1 A_5 A_6 A_7$ so that P_2 belongs to M and is different from P_1. Then, there are at least 5 points in $\{A_1, A_2, A_3, A_4, A_5, A_6, A_7\}$ which do not lie on line $P_1 P_2$. By the Pigeon Hole Principle, there exist at least 3 points on one side of line $P_1 P_2$, and these 3 points together with P_1 and P_2 constitute a convex pentagon which contains at least one point P_3 belonging to M.

Now, we have three lines $P_1 P_2$, $P_2 P_3$ and $P_3 P_1$, which form a triangle $\triangle P_1 P_2 P_3$. Let π_1 denote the half-plane on one side of line $P_1 P_2$ which is opposite to $\triangle P_1 P_2 P_3$ and contains no points on $P_1 P_2$. In a similar way, we define π_2 and π_3. Areas π_1, π_2 and π_3 cover the entire plane except $\triangle P_1 P_2 P_3$. By the Pigeon Hole Principle, there is one area of π_1, π_2 and π_3 which contains at least 3 points belonging to $\{A_1, A_2, A_3, A_4, A_5, A_6, A_7\}$, Without loss of generality, we assume that the area π_1 contains points A_1, A_2, A_3, then there exists one point P_4 belonging to M within the convex pentagon constituted by A_1, A_2, A_3, P_1 and P_2. So, $n \geqslant 11$.

Now, we give an example to illustrate that $n = 11$ is attainable. As seen in the figure, set M consists of integral points A_1, A_2, A_3, A_4, A_5, A_6, A_7, and four integral points within the heptagon $A_1 A_2 A_3 A_4 A_5 A_6 A_7$. Obviously, M satisfies Condition (1). We are going to

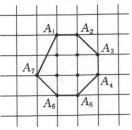

prove that M also satisfies Condition (2).

By reduction to absurdity, assume that there is a convex pentagon with its vertices belonging to M which contains no point of M in its interior. Then among such pentagons there must be one, denoted by $ABCDE$, which has the least area, since the value of the area of a polygon with integral vertices is always in the form of $\dfrac{n}{2}$ ($n \in \mathbf{N}$).

There are only 4 cases concerning the odd/even property of the xy-coordinate of a integral point: (odd, even), (even, odd), (odd, odd), (even, even). So there must be two vertices among A, B, C, D, E which have the same odd/even property, and the midpoint of the segment formed by these two vertices, say P, is also an integral point and belongs to M. By definition, P is not in the interior of pentagon $ABCDE$, then it must be on one side of the pentagon. Assume that P is on the side AB, then it must be the midpoint of AB, and $PBCDE$ is a convex pentagon with strictly less area than that of $ABCDE$.

So, the minimum value of n is 11.

Second Day
8:30 – 13:00   January 9, 2004

⬤ For a given real number a and a positive integer n, prove that:

(1) there exists exactly one sequence of real numbers x_0, x_1, $\cdots$, x_n, x_{n+1}, such that

$$\begin{cases} x_0 = x_{n+1} = 0, \\ \dfrac{1}{2}(x_{i+1} + x_{i-1}) = x_i + x_i^3 - a^3, \ i = 1, 2, \cdots, n; \end{cases}$$

(2) the sequence x_0, x_1, $\cdots$, x_n, x_{n+1} in (1) satisfies $|x_i| \leqslant |a|$, $i = 0, 1, \cdots, n+1$. (posed by Liang Yingde)

Solution   (1) Proof of existence: From $x_{i+1} = 2x_i + 2x_i^3 - 2a^3 - x_{i-1}$, $i = 1, 2, \cdots, n$ and $x_0 = 0$, we get that x_i is a polynomial of x_1

with degree 3^{i-1} and real coefficients, for $1 \leqslant i \leqslant n+1$. Specifically, x_{n+1} is a polynomial of x_1 with degree 3^n and real coefficients. As 3^n is an odd number, there exists a real number x_1 such that $x_{n+1} = 0$. Then from this x_1 and $x_0 = 0$ we can calculate x_i. The sequence $x_0, x_1, \cdots,$ x_n, x_{n+1} obtained in this way satisfies the required condition.

Proof of uniqueness: Suppose there are two sequences $w_0,$ $w_1, \cdots, w_n, w_{n+1}$ and $v_0, v_1, \cdots, v_n, v_{n+1}$, both of which satisfy Condition (1). Then $\frac{1}{2}(w_{i+1} + w_{i-1}) = w_i + w_i^3 - a^3$ and $\frac{1}{2}(v_{i+1} + v_{i-1}) = v_i + v_i^3 - a^3$. Thus,

$$\frac{1}{2}(w_{i+1} - v_{i+1} + w_{i-1} - v_{i-1})$$

$$= (w_i - v_i)(1 + w_i^2 + w_i v_i + v_i^2).$$

Suppose $|w_{i_0} - v_{i_0}|$ is the greatest. Then

$$|w_{i_0} - v_{i_0}| \leqslant |w_{i_0} - v_{i_0}|(1 + w_{i_0}^2 + w_{i_0} v_{i_0} + v_{i_0}^2)$$

$$\leqslant \frac{1}{2}|w_{i_0+1} - v_{i_0+1}| + \frac{1}{2}|w_{i_0-1} - v_{i_0-1}|$$

$$\leqslant |w_{i_0} - v_{i_0}|.$$

So, either $|w_{i_0} - v_{i_0}| = 0$ or $(1 + w_{i_0}^2 + w_{i_0} v_{i_0} + v_{i_0}^2) = 0$, that is, either $|w_{i_0} - v_{i_0}| = 0$ or $w_{i_0}^2 + v_{i_0}^2 + (w_{i_0} + v_{i_0})^2 = 0$. However it must be $|w_{i_0} - v_{i_0}| = 0$. Since $|w_{i_0} - v_{i_0}|$ is the greatest. Then $|w_i - v_i| = 0$ for every $i = 1, 2, \cdots, n$. This completed the proof of (1).

(2) Suppose that $|x_{i_0}|$ is the greatest. Then we have

$$|x_{i_0}| + |x_{i_0}|^3 = |x_{i_0}|(1 + |x_{i_0}|^2)$$

$$= \left|\frac{1}{2}(x_{i_0+1} + x_{i_0-1}) + a^3\right|$$

$$\leqslant \frac{1}{2}|x_{i_0+1}| + \frac{1}{2}|x_{i_0-1}| + |a|^3$$

$$\leqslant |x_{i_0}| + |a|^3.$$

That means $|x_{i_0}| \leqslant |a|$. So $|x_i| \leqslant |a|$, $i = 0, 1, \cdots, n+1$.

⑤ For a given positive integer $n \geqslant 2$, suppose positive integers $a_i (i = 1, 2, \cdots, n)$ satisfy $a_1 < a_1 < \cdots < a_n$ and $\sum\limits_{i=1}^{n} \dfrac{1}{a_i} \leqslant 1$.

Prove that, for any real number x, the following inequality holds,

$$\left(\sum_{i=1}^{n} \frac{1}{a_i^2 + x^2} \right)^2 \leqslant \frac{1}{2} \cdot \frac{1}{a_1(a_1 - 1) + x^2}.$$

(posed by Li Shenghong)

Solution For $x^2 \geqslant a_1(a_1 - 1)$, from $\sum\limits_{i=1}^{n} \dfrac{1}{a_i} \leqslant 1$ we have

$$\left(\sum_{i=1}^{n} \frac{1}{a_i^2 + x^2} \right)^2 \leqslant \left(\sum_{i=1}^{n} \frac{1}{2a_i |x|} \right)^2 = \frac{1}{4x^2} \left(\sum_{i=1}^{n} \frac{1}{a_i} \right)^2$$

$$\leqslant \frac{1}{4x^2} \leqslant \frac{1}{2} \cdot \frac{1}{a_1(a_1 - 1) + x^2}.$$

For $x^2 < a_1(a_1 - 1)$, using the Cauchy Inequality, we have

$$\left(\sum_{i=1}^{n} \frac{1}{a_i^2 + x^2} \right)^2 \leqslant \left(\sum_{i=1}^{n} \frac{1}{a_i} \right) \left(\sum_{i=1}^{n} \frac{a_i}{(a_i^2 + x^2)^2} \right) \leqslant \sum_{i=1}^{n} \frac{a_i}{(a_i^2 + x^2)^2}.$$

Further, for positive integers $a_1 < a_1 < \cdots < a_n$, we have $a_{i+1} \geqslant a_i + 1$ and

$$\frac{2a_i}{(a_i^2 + x^2)^2} \leqslant \frac{2a_i}{\left(a_i^2 + x^2 + \dfrac{1}{4} \right)^2 - a_i^2}$$

$$= \frac{2a_i}{\left(\left(a_i - \dfrac{1}{2} \right)^2 + x^2 \right) \left(\left(a_i + \dfrac{1}{2} \right)^2 + x^2 \right)}$$

$$= \frac{1}{\left(a_i - \dfrac{1}{2}\right)^2 + x^2} - \frac{1}{\left(a_i + \dfrac{1}{2}\right)^2 + x^2}$$

$$\leqslant \frac{1}{\left(a_i - \dfrac{1}{2}\right)^2 + x^2} - \frac{1}{\left(a_{i+1} - \dfrac{1}{2}\right)^2 + x^2},$$

for $i = 1, 2, \cdots, n-1$. So

$$\sum_{i=1}^{n} \frac{a_i}{(a_i^2 + x^2)^2} \leqslant \frac{1}{2} \sum_{i=1}^{n} \left[\frac{1}{\left(a_i - \dfrac{1}{2}\right)^2 + x^2} - \frac{1}{\left(a_{i+1} - \dfrac{1}{2}\right)^2 + x^2} \right]$$

$$\leqslant \frac{1}{2} \cdot \frac{1}{\left(a_1 - \dfrac{1}{2}\right)^2 + x^2} \leqslant \frac{1}{2} \cdot \frac{1}{a_1(a_1 - 1) + x^2}.$$

     Prove that every positive integer n, except a finite number of them, can be represented as a sum of 2 004 positive integers: $n = a_1 + a_2 + \cdots + a_{2\,004}$, where $1 \leqslant a_1 < a_2 < \cdots < a_{2\,004}$, and $a_i \mid a_{i+1}$, $i = 1, 2, \cdots, 2\,003$. (posed by Chen Yonggao)

Solution We are going to prove a more general result: For any positive integer $r \geqslant 2$, there exists $N(r) \in \mathbf{N}$ such that for every $n \geqslant N(r)$, there are positive integers $a_1, a_2, \cdots, a_r$ satisfying

$$n = a_1 + a_2 + \cdots + a_r, \ 1 \leqslant a_1 < a_2 < \cdots$$

$$< a_r, \ a_i \mid a_{i+1}, \ i = 1, 2, \cdots, r-1.$$

     For $r = 2$, we have $n = 1 + n - 1$, then $N(2) = 3$.

     Suppose it is true for $r = k$, then for $r = k + 1$ let $N(k+1) = 4N(k)^3$. For any positive integer $n = 2^\alpha(2l+1) \geqslant N(k+1) = 4N(k)^3$, we have either $2^\alpha \geqslant 2N(k)^2$ or $2l+1 > 2N(k)$.

     If $2^\alpha \geqslant 2N(k)^2$, there exists an even positive integer $2t \leqslant \alpha$ such that $2^{2t} \geqslant N(k)^2$, then $2^t + 1 \geqslant N(k)$. By induction we get positive integers $b_1, b_2, \cdots, b_k$ such that

$$2^t + 1 = b_1 + b_2 + \cdots + b_k, \quad 1 \leqslant b_1 < b_2 < \cdots$$
$$< b_k, \quad b_i \mid b_{i+1}, \quad i = 1, 2, \cdots, k-1.$$

Then

$$2^\alpha = 2^{\alpha-2t} \cdot 2^{2t} = 2^{\alpha-2t}[1 + (2^t - 1)](2^t + 1)$$

$$= 2^{\alpha-2t} + 2^{\alpha-2t}(2^t - 1)b_1 + 2^{\alpha-2t}(2^t - 1)b_2$$

$$+ \cdots + 2^{\alpha-2t}(2^t - 1)b_k.$$

$$n = 2^{\alpha-2t}(2l + 1) + 2^{\alpha-2t}(2^t - 1)b_1(2l + 1)$$

$$+ \cdots + 2^{\alpha-2t}(2^t - 1)b_k(2l + 1).$$

If $2l + 1 > 2N(k)$, then $l > N(k)$. By induction there exist positive integers $c_1, c_2, \cdots, c_k$ such that $l = c_1 + c_2 + \cdots + c_k$, $1 \leqslant c_1 < c_2 < \cdots < c_k$, $c_i \mid c_{i+1}$, $i = 1, 2, \cdots, k-1$. Then

$$n = 2^\alpha + 2^{\alpha+1}c_1 + 2^{\alpha+1}c_2 + \cdots + 2^{\alpha+1}c_k,$$

which is what we want. This completes the proof.

2005 (Zhengzhou, Henan)

2005 China Mathematical Olympiad and 20th Mathematics Winter Camp was held on January 20 – 25 in Zhengzhou, Henan Province, and was hosted by Olympiad Committee of CMS, the editorial deboard of 《Zhong Xue Sheng Shu Li Hua (中学生数理化)》 and Zhengzhou Foreign Language School.

The Competition Committee consisted of the following: Xiong Bin, Chen Yonggao, Leng Gangsong, Li Shenghong, Su Chun, Wang Jianwei, Wu Xihuan, Ye Zhonghao, Zhang Zhengjie, Zhu Huawei.

First Day
8:00 – 12:30 January 22, 2005

Let $\theta_i \in \left(-\dfrac{\pi}{2}, \dfrac{\pi}{2}\right)$, $i = 1, 2, 3, 4$. Prove that there exists $x \in$ **R** such that the following two inequalities

$$\cos^2 \theta_1 \cos^2 \theta_2 - (\sin \theta_1 \sin \theta_2 - x)^2 \geqslant 0, \qquad ①$$

$$\cos^2 \theta_3 \cos^2 \theta_4 - (\sin \theta_3 \sin \theta_4 - x)^2 \geqslant 0, \qquad ②$$

hold simultaneously if and only if

$$\sum_{i=1}^{4} \sin^2 \theta_i \leqslant 2\left(1 + \prod_{i=1}^{4} \sin \theta_i + \prod_{i=1}^{4} \cos \theta_i\right). \qquad ③$$

(posed by Li Shenghong)

Proof   Clearly, ① and ② are equivalent to

$$\sin \theta_1 \sin \theta_2 - \cos \theta_1 \cos \theta_2 \leqslant x \leqslant \sin \theta_1 \sin \theta_2 + \cos \theta_1 \cos \theta_2, \qquad ④$$

$$\sin \theta_3 \sin \theta_4 - \cos \theta_3 \cos \theta_4 \leqslant x \leqslant \sin \theta_3 \sin \theta_4 + \cos \theta_3 \cos \theta_4, \qquad ⑤$$

respectively. It is easy to show that there exists $x \in$ **R** such that ④ and ⑤ hold simultaneously if and only if

$$\sin \theta_1 \sin \theta_2 + \cos \theta_1 \cos \theta_2 - \sin \theta_3 \sin \theta_4 + \cos \theta_3 \cos \theta_4 \geqslant 0, \qquad ⑥$$

$$\sin \theta_3 \sin \theta_4 + \cos \theta_3 \cos \theta_4 - \sin \theta_1 \sin \theta_2 + \cos \theta_1 \cos \theta_2 \geqslant 0. \qquad ⑦$$

On the other hand, using $\sin^2 \alpha = 1 - \cos^2 \alpha$, we can simplify ③ into

$$\cos^2 \theta_1 \cos^2 \theta_2 + 2\cos \theta_1 \cos \theta_2 \cos \theta_3 \cos \theta_4 + \cos^2 \theta_3 \cos^2 \theta_4$$

$$- \sin^2 \theta_1 \sin^2 \theta_2 + 2\sin \theta_1 \sin \theta_2 \sin \theta_3 \sin \theta_4 - \sin^2 \theta_3 \sin^2 \theta_4 \geqslant 0,$$

or

$$(\cos \theta_1 \cos \theta_2 + \cos \theta_3 \cos \theta_4)^2 - (\sin \theta_1 \sin \theta_2 - \sin \theta_3 \sin \theta_4)^2 \geqslant 0.$$

That is,

$$(\sin \theta_1 \sin \theta_2 + \cos \theta_1 \cos \theta_2 - \sin \theta_3 \sin \theta_4 + \cos \theta_3 \cos \theta_4) \cdot$$

$(\sin \theta_3 \sin \theta_4 + \cos \theta_3 \cos \theta_4 - \sin \theta_1 \sin \theta_2 + \cos \theta_1 \cos \theta_2) \geqslant 0.$ ⑧

If there exists $x \in \mathbf{R}$ such that ④ and ⑤ hold simultaneously, then from ⑥ and ⑦ we can get ⑧ immediately. It follows that ③ holds.

Conversely, if ③ holds, or equivalently ⑧ holds, but ⑥ and ⑦ do not hold, then we have

$$\sin \theta_1 \sin \theta_2 + \cos \theta_1 \cos \theta_2 - \sin \theta_3 \sin \theta_4 + \cos \theta_3 \cos \theta_4 < 0,$$

and

$$\sin \theta_3 \sin \theta_4 + \cos \theta_3 \cos \theta_4 - \sin \theta_1 \sin \theta_2 + \cos \theta_1 \cos \theta_2 < 0.$$

Adding the last two equations, we obtain

$$2(\cos \theta_1 \cos \theta_2 + \cos \theta_3 \cos \theta_4) < 0,$$

which contradicts to the fact $\theta_i \in \left(-\dfrac{\pi}{2}, \dfrac{\pi}{2}\right)$, $i = 1, 2, 3, 4$. So ⑥ and ⑦ hold simultaneously. Hence there exists $x \in \mathbf{R}$ such that ④ and ⑤ hold simultaneously.

A circle intersects sides BC, CA, AB of ABC at two points for each side in the following order: $\{D_1, D_2\}$, $\{E_1, E_2\}$ and $\{F_1, F_2\}$. Line segments $D_1 E_1$ and $D_2 E_2$ intersect at point L, $E_1 F_1$ and $E_2 D_2$ intersect at point M, $F_1 D_1$ and $F_2 E_2$ intersect at point N. Prove that AL, BM and CN are concurrent. (posed by Ye Zhonghao)

Proof   Through point L draw perpendicular lines to AB and to AC, the feet are L' and L'' respectively. Let $\angle LAB = \alpha_1$, $\angle LAC = \alpha_2$, $\angle LF_2 A = \alpha_3$, and $\angle LE_1 A = a_4$. We have

$$\frac{\sin \alpha_1}{\sin \alpha_2} = \frac{LL'}{LL''} = \frac{LF_2 \sin \alpha_3}{LE_1 \sin \alpha_4}.$$ ①

Draw line segments $D_1 F_2$ and $D_2 E_1$ (see Figure 1). Since $\triangle L D_1 F_2 \backsim \triangle L D_2 E_1$ we get

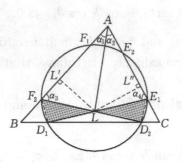

Figure 1

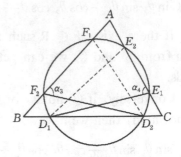

Figure 2

$$\frac{LF_2}{LE_1} = \frac{D_1 F_2}{D_2 E_1}. \tag{②}$$

Draw line segments $D_2 F_1$ and $D_1 E_2$ (see Figure 2). By using the sine rule we obtain

$$\frac{\sin \alpha_3}{\sin \alpha_4} = \frac{D_2 F_1}{D_1 E_2}. \tag{③}$$

Substituting ② and ③ into ①, we have

$$\frac{\sin \alpha_1}{\sin \alpha_2} = \frac{D_1 F_2}{D_2 E_1} \cdot \frac{D_2 F_1}{D_1 E_2}. \tag{④}$$

Similarly, write $\angle BMC = \beta_1$, $\angle MBA = \beta_2$, $\angle NCA = \gamma_1$, $\angle NCB = \gamma_2$, and we get

$$\frac{\sin \beta_1}{\sin \beta_2} = \frac{E_1 D_2}{E_2 F_1} \cdot \frac{E_2 D_1}{E_1 F_2}, \tag{⑤}$$

$$\frac{\sin \gamma_1}{\sin \gamma_2} = \frac{F_1 E_2}{F_2 D_1} \cdot \frac{F_2 E_1}{F_1 D_2}. \tag{⑥}$$

Multipling ④, ⑤ and ⑥, we obtain

$$\frac{\sin \alpha_1}{\sin \alpha_2} \cdot \frac{\sin \beta_1}{\sin \beta_2} \cdot \frac{\sin \gamma_1}{\sin \gamma_2} = 1.$$

Finally, according to the inverse of Ceva's Theorem, we know A L, BM and CN have a common point.

As seen in Figure 3, a circular pool is divided into $2n$ ($n \geqslant 5$) 'grids'. Two grids are said to be neighbors if they have a common side or an arc. It is easy to see that every grid has three neighbors.

$4n+1$ frogs jump into the pool. It is difficult for frogs to be quiet together. Whenever there is a grid where at least three frogs live together, sooner or later there must be three frogs in the grid jumping out simultaneously into the three neighboring grids respectively. Show that after a few times, the distribution of frogs in the pool will become uniform.

Here "uniform" means, for every grid of the pool, either the grid itself or each of the three neighboring grids has at least one frog. (posed by Su Chun)

Proof We call an event that there are three frogs in the same grid simultaneously jumping into three different neighboring grids " *an eruption* ". A grid is said to be " in equilibrium" if there are frogs in it or there are frogs in all its three neighbors.

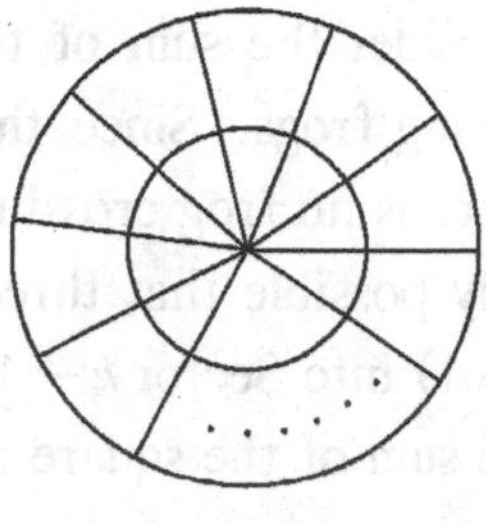

It is easy to see that a grid will be in

Figure 3

equilibrium after a frog jumps into the grid. Infact, frogs in this grid never move if there is no eruption. So, the grid is in equilibrium. If an eruption occurs, then there is at least one frog in each of its three neighbors. Further, it will be in equilibrium provided no eruption in each of its three neighbors. However, no matter which neighbor erupts, there will be frogs jumping into the grid, and there are frogs in it, so it will remain in equilibrium.

According to the above discussion, it suffices to prove that for any grid sooner or latter there is a frog jumping into it.

Given any grid, say Grid A. We call the sector in which the grid is situated Sector 1, and Grid B, for the other grid in Sector 1 (as shown in Figure 4). We have to prove that sooner or latter there are frogs jumping into Grid A.

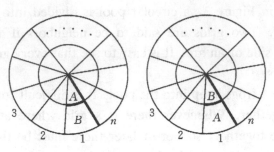

Figure 4

Number the remaining sectors 2 to n in the clockwise direction. At first we show that sooner or latter there are frogs jumping into Sector 1. .Suppose that in Sector 1 there are no frogs coming. Then there are no frogs crossing the wall between Sector 1 and Sector n. Consider the sum of the squares of the labeling numbers of sectors having frogs. Since there is no frog entering Sector 1 (especially, there is no frog crossing the wall between Sector 1 and Sector n), it is only possible that three frogs jumping from some Sector k ($3 \leqslant k \leqslant n-1$) into Sector $k-1$, k, and $k+1$, respectively. So the change in the sum of the square numbers is

$$(k-1)^2 + k^2 + (k+1)^2 - 3k^2 = 2.$$

That is increasing by 2. On the one hand, since frogs cannot stop jumping because there is at least one grid in which there are at least three frogs, hence the increasing of the sum does not stop. On the other hand, the sum cannot increase forever (it cannot be greater than $(4n+1)n^2$), a contradiction. Therefore, sooner or latter there are frogs that cross the wall between sector 1 and sector n, and enter Sector 1.

Next, we prove that sooner or latter there are three frogs jumping into Sector 1. If there are at most two frogs jumping into Sector 1, then they do not jumpout, and the above sum of the square numbers decreases at most twice (it happens only when two frogs cross the wall between Sector 1 and Sector n). And after that the sum increases continuously, again a contradiction. Hence, sooner or latter there are three frogs that jump into sector 1.

If among these three frogs some are in Grid A, then there are already frogs jumping into Grid A. Otherwise, these three frogs are all in Grid B, thus an eruption happens in Grid B and there is one frog that jumps into Grid A.

Second Day
8:00 – 12:30   January 23, 2005

Let $\{a_n\}$ be a sequence such that $a_1 = \dfrac{21}{16}$ and

$$2a_n - 3a_{n-1} = \frac{3}{2^{n+1}}, \ n \geqslant 2. \tag{①}$$

Let m be a positive integer and $m \geqslant 2$. Prove that for $n \leqslant m$,

$$\left[a_n + \frac{3}{2^{n+3}} \right]^{\frac{1}{m}} \left(m - \left(\frac{2}{3} \right)^{\frac{n(m-1)}{m}} \right) < \frac{m^2 - 1}{m - n + 1}. \tag{②}$$

(posed by Zhu Huawei)

Proof   By Equation ①, we have

$$2^n a_n = 3 \cdot 2^{n-1} a_{n-1} + \frac{3}{4}.$$

Set $b_n = 2^n a_n$, $n = 1, 2, \cdots$, then

$$b_n = 3b_{n-1} + \frac{3}{4}, \ b_n + \frac{3}{8} = 3\left(b_{n-1} + \frac{3}{8} \right).$$

Since $b_1 = 2a_1 = \dfrac{21}{8}$,

$$b_n + \frac{3}{8} = 3^{n-1}\left(b_1 + \frac{3}{8} \right) = 3^n,$$

it follows that

$$a_n = \left(\frac{3}{2}\right)^n - \frac{3}{2^{n+3}}.$$

Therefore, in order to prove Equation ②, it suffices to prove that

$$\left(\frac{3}{2}\right)^{\frac{n}{m}}\left(m - \left(\frac{2}{3}\right)^{\frac{n(m-1)}{m}}\right) < \frac{m^2 - 1}{m - n + 1},$$

or equivalently,

$$\left(1 - \frac{n}{m+1}\right)\left(\frac{3}{2}\right)^{\frac{n}{m}}\left(m - \left(\frac{2}{3}\right)^{\frac{n(m-1)}{m}}\right) < m - 1. \qquad ③$$

At first, we estimate the upper bound of $1 - \dfrac{n}{m+1}$. By using Bernoulli's inequality, we get

$$1 - \frac{n}{m+1} < \left(1 - \frac{1}{m+1}\right)^n,$$

so that

$$\left(1 - \frac{n}{m+1}\right)^m < \left(1 - \frac{1}{m+1}\right)^{nm} = \left(\frac{m}{m+1}\right)^{nm} = \left[\frac{1}{\left(1 + \frac{1}{m}\right)^m}\right]^n.$$

(Note: By the mean inequality, we can also have the same result:

$$\left(1 - \frac{n}{m+1}\right)^m = \left(1 - \frac{n}{m+1}\right)^m \cdot \underbrace{1 \cdot 1 \cdot \cdots \cdot 1}_{\text{number } mn-m \text{ of } 1}$$

$$< \left[\frac{m\left(1 - \frac{n}{m+1}\right) + mn - m}{mn}\right]^{mn}$$

$$= \left(\frac{m}{m+1}\right)^{nm}.$$

Since $m \geqslant 2$, in view of the binomial formula, we obtain

$$\left(1+\frac{1}{mm}\right)^m \geqslant 1+C_m^1 \cdot \frac{1}{m}+C_m^2 \cdot \frac{1}{m^2}=\frac{5}{2}-\frac{1}{2m} \geqslant \frac{9}{4}.$$

It follows that

$$\left(1-\frac{n}{m+1}\right)^m < \left(\frac{4}{9}\right)^n,$$

or

$$1-\frac{n}{m+1} < \left(\frac{2}{3}\right)^{\frac{2n}{m}}.$$

Hence, if we want to prove Equation ③, we only need to prove that

$$\left(\frac{2}{3}\right)^{\frac{2n}{m}} \cdot \left(\frac{3}{2}\right)^{\frac{n}{m}}\left(m-\left(\frac{2}{3}\right)^{\frac{n(m-1)}{m}}\right)<m-1,$$

that is,

$$\left(\frac{2}{3}\right)^{\frac{n}{m}}\left(m-\left(\frac{2}{3}\right)^{\frac{n(m-1)}{m}}\right)<m-1. \qquad ④$$

Set $\left(\frac{2}{3}\right)^{\frac{n}{m}}=t$, then $0<t<1$, and Equation ④ now becomes

$$t(m-t^{m-1})<m-1,$$

or

$$(t-1)[m-(t^{m-1}+t^{m-2}+\cdots+1)]<0.$$

The above inequality clearly holds, so does the initial inequality.

🔷 There are 5 points in a rectangle $ABCD$ (including its boundary) with unit area such that any three of them are not collinear. Find the minimun number of triangles with areas no more than $\frac{1}{4}$ and vertexes chosen from these 5 points. (posed by Leng Gangsong)

Solution　At first, we give a lemma without proof.

Lemma *The area of a triangle inscribed in a rectangle is no more than half of the area of the rectangle.*

In the rectangle $ABCD$ if there are 3 points such that the area of the triangle with these points as the vertices is no more than $\dfrac{1}{4}$, then these 3 points are called *a good triple or good*.

Denote E, F, H and G the midpoints of AB, CD, BC and AD, respectively, and O the intersection of line segments EF and GH. EF and GH divide the rectangle $ABCD$ into 4 small rectangles, it follows that there exists certain small rectangle, say $AEOG$, in which there are at least two points (say, M and N) out of those 5 points, see Figure 5.

(1) If there is no more than one given point in the rectangle $OHCF$, consider any given point X which is different from M and N and not in the rectangle $OHCF$. It is easy to verify that triple (M, N, X) is either in rectangle $ABHG$, or in rectangle $AEFD$. By the above lemma (M, N, X) is good. Since there are at lease two such points X, we have at least two such good triples.

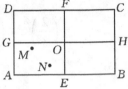

Figure 5

(2) If there exist at least two given points in rectangle $OHCF$, we suppose that P and Q are the given points in rectangle $OHCF$. Consider the final given point R. If R is in the rectangle $OFDG$, then (M, N, R) is in the rectangle $AEFD$, and (P, Q, R) is in the rectangle $GHCD$. So they are all good. It follows that there are at least two good triples. Similarly, when point R is in the rectangle $EBHO$ there are at least two good triples too. If point R is in the rectangle $OHCF$ or the rectangle $AEOG$, suppose that R is in the rectangle $OHCF$. Consider the smallest convex polygon containing the 5 points M, N, P, Q and R. The polygon must be contained in the convex hexagon $AEHCFG$ (see Figure 6). But the area

$$S_{AEHCFG} = 1 - \frac{1}{8} - \frac{1}{8} = \frac{3}{4}.$$

We divide the case into three sub-cases as follows.

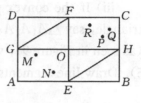

Figure 6

i) Suppose the convex polygon generated by M, N, P, Q and R is a convex pentagon, (no loss of generality) say $MNPQR$ (as seen in Figure 7). In this case

$$S_{\triangle MQR} + S_{\triangle MNQ} + S_{\triangle NPQ} \leqslant \frac{3}{4},$$

it follows that there is at least one good triple among (M, Q, R), (M, N, Q) and (N, P, Q). Moreover, (P, Q, R) is clearly good for it is in the rectangle $OHCF$. Thus there are at least two good triples.

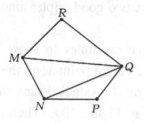

Figure 7

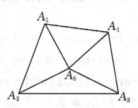

Figure 8

ii) If the convex polygon generated by M, N, P, Q and R is a convex quadrilateral, say $A_1 A_2 A_3 A_4$, and the fifth point is A_5 (as seen in Figure 8) where $A_i \in \{M, N, P, Q, R\}$ $(i = 1, 2, 3, 4, 5)$. Draw line segments $A_5 A_i (i = 1, 2, 3, 4)$, then

$$S_{\triangle A_1 A_2 A_5} + S_{\triangle A_2 A_3 A_5} + S_{\triangle A_3 A_4 A_5} + S_{\triangle A_4 A_1 A_5} = S_{\triangle A_1 A_2 A_3 A_4} \leqslant \frac{3}{4}.$$

Therefore, there are at least two good triples among (A_1, A_2, A_5), (A_2, A_3, A_5), (A_3, A_4, A_5) and (A_1, A_4, A_5).

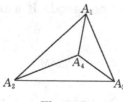

Figure 9

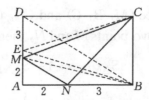

Figure 10

iii) If the convex polygon generated by M, N, P, Q and R is a triangle, say $\triangle A_1 A_2 A_3$, and the remaining two points are A_4 and A_5 (as seen in Figure 9), where $A_i \in \{M, N, P, Q, R\}$ ($i = 1, 2, 3, 4, 5$). Draw line segments $A_4 A_i$ ($i = 1, 2, 3$), then

$$S_{\triangle A_1 A_2 A_4} + S_{\triangle A_2 A_3 A_4} + S_{\triangle A_3 A_1 A_4} = S_{\triangle A_1 A_2 A_3} \leqslant \frac{3}{4}.$$

Therefore, there is at least one good triple among (A_1, A_2, A_4), (A_2, A_3, A_4) and (A_1, A_3, A_4). Similarly, A_5 with two of A_1, A_2 and A_3 constitutes a good triple. Consequently, in this case there are at least two good triples.

Thus, in any case there are at least two good triples among these 5 points.

In the following we will give some examples to show that the number of good triples may be just two. Pick a point M on the side AD of the rectangle $ABCD$ and a point N on the side AB such that AN : $NB = AM$: $MD = 2$: 3 (as shown in Figure 10). Then among 5 points M, N, B, C and D there are just two good triples. In fact, (B, C, D) is clearly not good. Suppose that a triple containing exactly one of two points M and N, say M. Let E be the midpoint of AD, then

$$S_{\triangle MBD} > S_{\triangle EBD} = \frac{1}{4}.$$

It follows that (M, B, D) is not good. Thus

$$S_{\triangle MBC} = \frac{1}{2}, \quad S_{\triangle MCD} > S_{\triangle ECD} = \frac{1}{4}.$$

Hence (M, B, C) and (M, C, D) are not good. If a tripe does contain two points M and N, then

$$S_{\triangle MNC} = 1 - S_{\triangle NBC} - S_{\triangle MCD} - S_{\triangle AMN}$$

$$= 1 - \frac{3}{5} S_{\triangle ABC} - \frac{3}{5} S_{\triangle ACD} - \frac{4}{25} S_{\triangle ABD}$$

$$=1-\frac{3}{10}-\frac{3}{10}-\frac{2}{25}=\frac{8}{25}>\frac{1}{4}.$$

So (M, N, C) is not good. But $S_{\triangle MNB} = S_{\triangle MND} = \frac{1}{5} < \frac{1}{4}$, thus among the triples there are only two good triples (M, N, B) and (M, N, D).

Consequently, the minimun number of triangles with area not great than $\frac{1}{4}$ is 2.

Find all non-negative integer solutions (x, y, z, w) of the following equation

$$2^x \cdot 3^y - 5^z \cdot 7^w = 1.$$

(posed by Chen Yonggao)

Solution   Since $5^z \cdot 7^w + 1$ is even, we have $x \geqslant 1$.

Case 1: $y = 0$. The equation to be solved becomes

$$2^x - 5^z \cdot 7^w = 1.$$

If $z \neq 0$, then $2^x \equiv 1 \pmod 5$. It follows that $4 \mid x$. Thus $3 \mid 2^x - 1$, which contradicts to $2^x - 5^z \cdot 7^w = 1$.

If $z = 0$, then

$$2^x - 7^w = 1.$$

When $x = 1, 2, 3$, a direct computation shows that $(x, w) = (1, 0), (3, 1)$ are the solutions.

When $x \geqslant 4$, $7^w \equiv -1 \pmod{16}$. By direct computation we know that this is impossible.

Consequently, when $y = 0$ all non-negative integer solutions of the equation are

$$(x, y, z, w) = (1, 0, 0, 0), (3, 0, 0, 1).$$

Case 2: $y > 0$ and $x = 1$. Thus the equation to be solved becomes

$$2 \cdot 3^y - 5^z \cdot 7^w = 1.$$

Hence $-5^z \cdot 7^w \equiv 1 \pmod 3$, i.e., $(-1)^z \equiv -1 \pmod 3$. It follows that z is odd.

Thus

$$2 \cdot 3^y \equiv 1 \pmod 5.$$

So $y \equiv 1 \pmod 4$.

When $w \neq 0$, we have $2 \cdot 3^y \equiv 1 \pmod 7$. Thus $y \equiv 4 \pmod 6$, which contradicts to the fact $y \equiv 1 \pmod 4$. Hence $w = 0$ and

$$2 \cdot 3^y - 5^z = 1.$$

When $y = 1$, we have $z = 1$. If $y \geqslant 2$, then $5^z \equiv -1 \pmod 9$, which implies $z \equiv 3 \pmod 6$. Thus $5^3 + 1 \mid 5^z + 1$, so $7 \mid 5^z + 1$, which contradicts to $5^z + 1 = 2 \cdot 3^y$. Hence in this case we have only one solution

$$(x, y, z, w) = (1, 1, 1, 0).$$

Case 3: $y > 0$ and $x \geqslant 2$. Thus

$$5^z \cdot 7^w \equiv -1 \pmod 4, \text{ and } 5^z \cdot 7^w \equiv -1 \pmod 3.$$

That is,

$$(-1)^w \equiv -1 \pmod 4, \text{ and } (-1)^z \equiv -1 \pmod 3.$$

Thus z and w are odd. It follows that

$$2^x \cdot 3^y = 5^z \cdot 7^w + 1 \equiv 35 + 1 \equiv 4 \pmod 8.$$

Hence, $x = 2$, and

$$4 \cdot 3^y - 5^z \cdot 7^w = 1 \text{ (where } z \text{ and } w \text{ are odd).} \qquad ①$$

Thus,

$$4 \cdot 3^y \equiv 1 \pmod 5, \text{ and } 4 \cdot 3^y \equiv 1 \pmod 7.$$

From the above two congruencies we have $y \equiv 2 \pmod{12}$.
Set $y = 12m + 2$, $m \geqslant 0$, then

$$5^z \cdot 7^w = 4 \cdot 3^y - 1 = (2 \cdot 3^{6m+1} - 1)(2 \cdot 3^{6m+1} + 1).$$

Since

$$2 \cdot 3^{6m+1} + 1 \equiv 6 \cdot 2^{3m} + 1 \equiv 6 + 1 \equiv 0 \pmod{7},$$

and $(2 \cdot 3^{6m+1} - 1, 2 \cdot 3^{6m+1} + 1) = 1$, so $5 \mid 2 \cdot 3^{6m+1} - 1$. Thus

$$2 \cdot 3^{6m+1} - 1 = 5^z, \qquad \textcircled{2}$$
$$2 \cdot 3^{6m+1} + 1 = 7^w.$$

If $m \geqslant 1$, by Equation ② we have $5^z \equiv -1 \pmod 9$, and from Case 2 we know that this is impossible.

If $m = 0$, then $y = 2$, $z = 1$ and $w = 1$. Thus in this case, we have only one solution

$$(x, y, z, w) = (2, 2, 1, 1).$$

Consequently, all non-negative integer solutions are

$$(x, y, z, w) = (1, 0, 0, 0), (3, 0, 0, 1),$$
$$(1, 1, 1, 0), (2, 2, 1, 1).$$

2006 (Fuzhou, Fujian)

2006 China Mathematical Olympiad and 21th Mathematics Winter Camp was held on January 10 – 15 in Fuzhou, Fujian Province, and was hosted by Olympiad Committee of CMS and Fuzhou No. 1 middle school.

The Competition Committee consisted of the following: Li Shenghong, Li Yonggao, Xiong Bin, Leng Gangsong, Li Weigu, Wang Jianwei, Zhu Huawei, Lin Chang, Luo Wei, Ji Chungang.

First Day
8:00 – 12:30 January 12, 2006

Suppose that the real numbers a_1, a_2, $\cdots$, a_n satisfy $a_1 + a_2 + \cdots$

$+a_n = 0.$ **Prove**

$$\max_{1 \leqslant i \leqslant n} a_i^2 \leqslant \frac{n}{3} \sum_{i=1}^{n-1} (a_i - a_{i+1})^2. \text{ (posed by Zhu Huawei)}$$

Solution   It is sufficient to prove that: For every $k \in \{1, 2, \cdots, n\}$, we have

$$a_k^2 \leqslant \frac{n}{3} \sum_{i=1}^{n-1} (a_i - a_{i+1})^2.$$

Let $d_k = a_k - a_{k+1}$, $k = 1, 2, \cdots, n-1$, then

$$a_k = a_k,$$

$$a_{k+1} = a_k - d_k, \ a_{k+2} = a_k - d_k - d_{k+1}, \cdots,$$

$$a_n = a_k - d_k - d_{k+1} - \cdots - d_{n-1},$$

$$a_{k-1} = a_k + d_{k-1}, \ a_{k-2} = a_k + d_{k-1} + d_{k-2}, \cdots,$$

$$a_1 = a_k + d_{k-1} + d_{k-2} + \cdots + d_1.$$

Summing all equalities, and using $a_1 + a_2 + \cdots + a_n = 0$, we get

$$na_k - (n-k)d_k - (n-k-1)d_{k+1} - \cdots - d_{n-1}$$

$$+ (k-1)d_{k-1} + (k-2)d_{k-2} + \cdots + d_1 = 0.$$

Then in view of Cauchy's inequality

$$(na_k)^2 = ((n-k)d_k + (n-k-1)d_{k+1} + \cdots + d_{n-1}$$

$$- (k-1)d_{k-1} - (k-2)d_{k-2} - \cdots - d_1)^2$$

$$\leqslant \left(\sum_{i=1}^{k-1} i^2 + \sum_{i=1}^{n-k} i^2 \right) \left(\sum_{i=1}^{n-1} d_i^2 \right)$$

$$\leqslant \left(\sum_{i=1}^{n-1} i^2 \right) \left(\sum_{i=1}^{n-1} d_i^2 \right) = \frac{n(n-1)(2n-1)}{6} \left(\sum_{i=1}^{n-1} d_i^2 \right)$$

$$< \frac{n^3}{3} \left(\sum_{i=1}^{n-1} d_i^2 \right).$$

So, $a_k^2 \leqslant \dfrac{n}{3} \displaystyle\sum_{i=1}^{n-1} (a_i - a_{i+1})^2.$

🔷 Suppose that positive integers a_1, a_2, $\cdots$, $a_{2\,006}$ (some of them may be equal) satisfy the condition: any two of $\dfrac{a_1}{a_2}$, $\dfrac{a_2}{a_3}$, $\cdots$, $\dfrac{a_{2\,005}}{a_{2\,006}}$ are unequal. At least how many different numbers are there in $\{a_1$, a_2, $\cdots$, $a_{2\,006}\}$? (posed by Chen Yonggao)

Solution The answer is: there are at least 46 different numbers in $\{a_1$, a_2, $\cdots$, $a_{2\,006}\}$.

With 45 different positive integers we can only get $45 \times 44 + 1 = 1\,981$ fractions. So there are more than 45 different numbers in $\{a_1$, a_2, $\cdots$, $a_{2\,006}\}$.

On the other hand, let p_1, p_2, $\cdots$, p_{46} be 46 different prime. Set a_1, a_2, $\cdots$, $a_{2\,006}$ to be:

$$p_1, \ p_1, \ p_2, \ p_1, \ p_3, \ p_2, \ p_3, \ p_1, \ p_4, \ p_3, \ p_4, \ p_2, \ p_4, \ p_1, \ \cdots,$$

$$p_1, \ p_k, \ p_{k-1}, \ p_k, \ p_{k-2}, \ p_k, \ \cdots, \ p_k, \ p_2, \ p_k, \ p_1, \ \cdots,$$

$$p_1, \ p_{45}, \ p_{44}, \ p_{45}, \ p_{43}, \ p_{45}, \ \cdots, \ p_{45}, \ p_2, \ p_{45}, \ p_1,$$

$$p_{46}, \ p_{45}, \ p_{46}, \ p_{44}, \ p_{46}, \ \cdots, \ p_{46}, \ p_{22}, \ p_{46}.$$

Then the 2 006 positive numbers satisfy that any two of $\dfrac{a_1}{a_2}$, $\dfrac{a_2}{a_3}$, $\cdots$, $\dfrac{a_{2\,005}}{a_{2\,006}}$ are unequal.

So the answer is 46.

🔷 Suppose positive integers m, n, k satisfy $mn = k^2 + k + 3$. Prove that at least one of the following Diophantine equations

$$x^2 + 11y^2 = 4m \text{ and } x^2 + 11y^2 = 4n$$

has a solution (x, y) with x, y being odd numbers. (posed by Li Weigu)

Solution First, we prove a lemma.

Lemma *The following Diophantine equation*

$$x^2 + 11y^2 = 4m \qquad \qquad ①$$

has a solution (x_0, y_0)*, such that either* x_0, y_0 *are odd numbers or* x_0, y_0 *are even numbers with* $x_0 \equiv (2k+1)y_0 \pmod{m}$. ②

Proof of the lemma: Consider the expression $x + (2k+1)y$, where x, y are integers, and $0 \leqslant x \leqslant 2\sqrt{m}$, $0 \leqslant y \leqslant \dfrac{\sqrt{m}}{2}$.

There are $([2\sqrt{m}]+1)\left[\left[\dfrac{\sqrt{m}}{2}\right]+1\right] > m$ such expressions. So there exist integers $x_1, x_2 \in [0, 2\sqrt{m}]$, $y_1, y_2 \in \left[0, \dfrac{\sqrt{m}}{2}\right]$, such that $(x_1, y_1) \neq (x_2, y_2)$, and

$$x_1 + (2k+1)y_1 \equiv x_2 + (2k+1)y_2 \pmod{m},$$

So $$x \equiv (2k+1)y \pmod{m}, \qquad \qquad ③$$

where $$x = x_1 - x_2, \quad y = y_2 - y_1.$$

This means

$$x^2 \equiv (2k+1)^2 y^2 \equiv -11y^2 \pmod{m},$$

that is, $$x^2 + y^2 = tm.$$

Since $|x| \leqslant 2\sqrt{m}$, $|y| \leqslant \dfrac{\sqrt{m}}{2}$, we have

$$x^2 + 11y^2 < 4m + \frac{11}{4}m < 7m.$$

So $$1 \leqslant t \leqslant 6.$$

As m is an odd number, obviously the equations $x^2 + 11y^2 = 2m$, and $x^2 + 11y^2 = 6m$ have no integer solution.

(1) If $x^2 + 11y^2 = m$, then $x_0 = 2x$, $y_0 = 2y$ is a solution of ① satisfying ②.

(2) If $x^2 + 11y^2 = 4m$, then $x_0 = x$, $y_0 = y$ is a solution of ①

satisfying ②.

(3) If $x^2 + 11y^2 = 3m$, then $(x \pm 11y)^2 + 11(x \mp y)^2 = 3^2 \cdot 4m$.

First, assume that $3 / m$. If $x \not\equiv 0 \pmod 3$, $y \not\equiv 0 \pmod 3$, and $x \not\equiv y \pmod 3$, then

$$x_0 = \frac{x - 11y}{3}, \quad y_0 = \frac{x + y}{3} \qquad \text{④}$$

is a solution of ① satisfying ②.

If $x \equiv y \not\equiv 0 \pmod 3$, then

$$x_0 = \frac{x + 11y}{3}, \quad y_0 = \frac{y - x}{3} \qquad \text{⑤}$$

is a solution of ① satisfying ②.

Now suppose $3/m$. Then ④ and ⑤ are still integer solutions of ①. If equation ① has an even integer solution $x_0 = 2x_1$, $y_0 = 2y_1$, then

$$x_1^2 + 11y_1^2 = m \Leftrightarrow 36m = (5x_1 \pm 11y_1)^2 + 11(5y_1 \mp x_1)^2.$$

Since one of x_1, y_1 is even and the other is odd, so $5x_1 \pm 11y_1$, $5y_1 \mp x_1$ are odd numbers.

If $x \equiv y \pmod 3$, then $x_0 = \dfrac{5x_1 - 11y_1}{3}$, $y_0 = \dfrac{5y_1 + x_1}{3}$ is a solution of ① satisfying ②.

If $x_1 \not\equiv y_1 \pmod 3$, then $x_0 = \dfrac{5x_1 + 11y_1}{3}$, $y_0 = \dfrac{5y_1 - x_1}{3}$ is a solution of ① satisfying ②.

(4) If $x^2 + 11y^2 = 5m$, then $5^2 \cdot 4m = (3x \mp 11y)^2 + 11(3y \pm x)^2$.
When $5 \nmid m$, if

$$x \equiv \pm 1 \pmod 5, \quad y \equiv \mp 2 \pmod 5,$$

$$\text{or } x \equiv \pm 2 \pmod 5, \quad y \equiv \pm 1 \pmod 5,$$

then

$$x_0 = \frac{3x - 11y}{5}, \quad y_0 = \frac{3y + x}{5} \qquad \text{⑥}$$

is a solution of ① satisfying ②.

If $x \equiv \pm 1 (\mod 5)$, $y \equiv \pm 2 (\mod 5)$, or $x \equiv \pm 2 (\mod 5)$, $y \equiv \mp 1 (\mod 5)$, then

$$x_0 = \frac{3x + 11y}{5}, \quad y_0 = \frac{3y - x}{5} \tag{⑦}$$

is a solution of ① satisfying ②.

When $5/m$, then ⑥ and ⑦ are still are integer solutions of ①. If Equation ① has an even integer solution $x_0 = 2x_1$, $y_0 = 2y_1$, then

$$x_1^2 + 11y_1^2 = m, \quad x_1 \not\equiv y_1 (\mod 2),$$

and we have $100m = (x_1 \mp 33y_1)^2 + 11(y_1 \pm 3x_1)^2$.

If $x_1 \equiv y_1 \equiv 0 (\mod 5)$, or $x_1 \equiv \pm 1 (\mod 5)$, $y_1 \equiv \pm 2 (\mod 5)$, or

$$x_1 \equiv \pm 2 (\mod 5), \quad y_1 \equiv \mp 1 (\mod 5),$$

then $x_0 = \dfrac{x_1 - 33y_1}{5}$, $y_0 = \dfrac{y_1 + 3x_1}{5}$ is a solution of ① satisfying ②.

If $x_1 \equiv \pm 1 (\mod 5)$, $y_1 \equiv \mp 2 (\mod 5)$, or $x_1 \equiv \pm 2 (\mod 5)$, $y_1 \equiv \pm 1 (\mod 5)$, then

$$x_0 = \frac{x_1 + 33y_1}{5}, \quad y_0 = \frac{y_1 - 33x_1}{5}$$

is a solution of ① satisfying ②.

The lemma is proved.

From the lemma, if ① has a solution (x, y) with x, y being odd numbers, then it has a solution (x_0, y_0) with x_0, y_0 being even numbers satifying ②.

Let $l = 2k + 1$, the solution of the quadratic equation

$$mx^2 + ly_0 x + ny_0^2 - 1 = 0. \tag{⑧}$$

is $x = \dfrac{-ly_0 \pm \sqrt{l^2 y_0^2 - 4mny_0^2 + 4m}}{2m} = \dfrac{-ly_0 \pm x_0}{2m}$. So Equation ⑧ has at least an integer solution x_1, i.e.

$$mx_1^2 + ly_0 x_1 + ny_0^2 - 1 = 0. \tag{9}$$

This indicates that x_1 is an odd number. Now, from ⑨, it follows that

$$(2ny_0 + lx_1)^2 + 11x_1^2 = 4n.$$

This means $x^2 + 11y^2 = 4n$ has a solution (x, y) with x, y being odd numbers, where $x = 2ny_0 + lx_1$, $y = x_1$.

Second Day

8:00 – 12:30　January 13, 2006

Let $\triangle ABC$ be a right-angled triangle with $\angle ACB = 90°$. The inscribed circle $\odot O$ of $\triangle ABC$ is tangent to BC, CA, AB at D, E, F respectively. AD intersects $\odot O$ at P, $\angle BPC = 90°$. Prove that $AE + AP = PD$. (posed by Xiong Bin)

Proof　Let $AE = AF = x$, BD $= BF = y$, $CD = CE = z$, AP $= m$, $PD = n$.

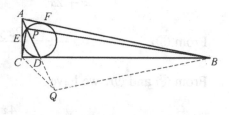

Since $\angle ACP + \angle PCB =$ $90° = \angle PBC + \angle PCB$, then $\angle ACP = \angle PBC$.

Let Q be a point on the line AD, such that $\angle AQC = \angle ACP = \angle PBC$. Then P, B, Q, C are concyclic.

Let $DQ = l$, then by the intersecting chord theorem and the tangent secant theorem, it follows that

$$yz = nl, \tag{1}$$

$$x^2 = m(m+n). \tag{2}$$

As $\triangle ACP \backsim \triangle AQC$, we have $\dfrac{AC}{AQ} = \dfrac{AP}{AC}$. So

$$(x+z)^2 = m(m+n+l). \tag{3}$$

In Rt $\triangle ACD$ and Rt $\triangle ACB$, using the Pythagorean theorem, we get

$$(x+z)^2 + z^2 = (m+n)^2, \qquad ④$$

$$(y+z)^2 + (z+x)^2 = (x+y)^2. \qquad ⑤$$

③ − ②:
$$z^2 + 2zx = ml, \qquad ⑥$$

① ÷ ⑥:
$$\frac{yz}{z^2 + 2zx} = \frac{n}{m}.$$

So
$$1 + \frac{yz}{z^2 + 2zx} = \frac{m+n}{m}. \qquad ⑦$$

② × ⑦, together with ①, we get $x^2 + \dfrac{x^2 yz}{z^2 + 2zx} = (m+n)^2 = (x+z)^2 + z^2$, i.e.

$$\frac{x^2 y}{z + 2x} = 2z(x+z). \qquad ⑧$$

From ⑤,
$$x + z = \frac{2xy}{y+z}. \qquad ⑨$$

From ⑧ and ⑨, we have

$$\frac{x}{z + 2x} = \frac{4z}{y+z}. \qquad ⑩$$

Also from ⑤, $y + z = \dfrac{2xz}{x-z}$. Substituting into ⑩ and eliminating $y + z$, we abtain

$$3x^2 - 2xz - 2z^2 = 0,$$

giving the solution $x = \dfrac{\sqrt{7}+1}{3} z$, $y = (2\sqrt{7}+5)z$. Together with ④, it implies

$$m + n = \frac{2(\sqrt{7}+1)}{3} z.$$

From ②, $m = \dfrac{x^2}{m+n} = \dfrac{\sqrt{7}+1}{6}z$, $n = \dfrac{\sqrt{7}+1}{2}z$. So $x+m=n$, i.e. $AE + AP = PD$.

⑤ A sequence of real numbers $\{a_n\}$ satisfies the condition that $a_1 = \dfrac{1}{2}$, $a_{k+1} = -a_k + \dfrac{1}{2-a_k}$, $k = 1, 2, \cdots$.

Prove the following inequality:

$$\left(\frac{n}{2(a_1 + a_2 + \cdots + a_n)} - 1\right)^n \leqslant \left[\frac{a_1 + a_2 + \cdots + a_n}{n}\right]^n$$

$$\left(\frac{1}{a_1} - 1\right)\left(\frac{1}{a_2} - 1\right)\cdots\left(\frac{1}{a_n} - 1\right).$$

(posed by Li Shenghong)

Proof First, we use induction to prove that $0 < a_n \leqslant \dfrac{1}{2}$, $n = 1, 2, \cdots$.

When $n = 1$, it is obvious.

Now suppose it is true for n $(n \geqslant 1)$, i.e. $0 < a_n \leqslant \dfrac{1}{2}$.

Let $f(x) = -x + \dfrac{1}{2-x}$ for $x \in \left[0, \dfrac{1}{2}\right]$. Then $f(x)$ is a decreasing function. So

$$a_{n+1} = f(a_n) \leqslant f(0) = \frac{1}{2},$$

$$a_{n+1} = f(a_n) \geqslant f\left(\frac{1}{2}\right) = \frac{1}{6} > 0,$$

i.e. it is true for $n+1$.

Back to the problem, it is sufficient to prove:

$$\left(\frac{n}{a_1 + a_2 + \cdots + a_n}\right)^n \left(\frac{n}{2(a_1 + a_2 + \cdots + a_n)} - 1\right)^n$$

$$\leqslant \left(\frac{1}{a_1} - 1\right)\left(\frac{1}{a_2} - 1\right)\cdots\left(\frac{1}{a_n} - 1\right).$$

Let $f(x) = \ln\left(\dfrac{1}{x} - 1\right)$ for $x \in \left(0, \dfrac{1}{2}\right)$. Then $f(x)$ is a concave function, i.e. for every $0 < x_1$, $x_2 < \dfrac{1}{2}$, we have $f\left(\dfrac{x_1 + x_2}{2}\right) \leqslant$

$\dfrac{f(x_1) + f(x_2)}{2}$.

In fact,

$$f\left(\frac{x_1 + x_2}{2}\right) \leqslant \frac{f(x_1) + f(x_2)}{2}$$

is equivalent to

$$\left(\frac{2}{x_1 + x_2} - 1\right)^2 \leqslant \left(\frac{1}{x_1} - 1\right)\left(\frac{1}{x_2} - 1\right),$$

$$\Leftrightarrow (x_1 - x_2)^2 \geqslant 0.$$

So $f(x)$ is a concave function.

Using Jenson's Inequality, It follows that

$$f\left(\frac{x_1 + x_2 + \cdots + x_n}{n}\right) \leqslant \frac{f(x_1) + f(x_2) + \cdots + f(x_n)}{n},$$

so $\left(\dfrac{n}{a_1 + a_2 + \cdots + a_n} - 1\right)^n \leqslant \left(\dfrac{1}{a_1} - 1\right)\left(\dfrac{1}{a_2} - 1\right)\cdots\left(\dfrac{1}{a_n} - 1\right).$

On the other hand, by using Cauchy's Inequality, we have

$$\sum_{i=1}^{n}(1 - a_i) = \sum_{i=1}^{n}\frac{1}{a_i + a_{i+1}} - n \geqslant \frac{n^2}{\sum_{i=1}^{n}(a_i + a_{i+1})} - n$$

$$= \frac{n^2}{a_{n+1} - a_1 + 2\sum_{i=1}^{n}a_i} - n$$

$$\geqslant \frac{n^2}{2\sum_{i=1}^{n}a_i} - n = n\left(\frac{n}{2\sum_{i=1}^{n}a_i} - 1\right).$$

This means $\dfrac{\displaystyle\sum_{i=1}^{n}(1-a_i)}{\displaystyle\sum_{i=1}^{n}a_i} \geqslant \dfrac{n}{\displaystyle\sum_{i=1}^{n}a_i}\left[\dfrac{n}{2\displaystyle\sum_{i=1}^{n}a_i}-1\right].$

So

$$\left(\frac{n}{a_1+a_2+\cdots+a_n}\right)^n\left(\frac{n}{2(a_1+a_2+\cdots+a_n)}-1\right)^n$$

$$\leqslant\left[\frac{(1-a_1)+(1-a_2)+\cdots+(1-a_n)}{a_1+a_2+\cdots+a_n}\right]^n$$

$$\leqslant\left(\frac{1}{a_1}-1\right)\left(\frac{1}{a_2}-1\right)\cdots\left(\frac{1}{a_n}-1\right).$$

Let X be a set of 56 elements. Find the least positive integer n such that for any 15 subsets of X, if the union of every 7 sets of these subsets contains at least n elements, then there exist 3 of the 15 subsets whose intersection is nonempty. (posed by Leng Gangsong)

Solution The minimum value of n is 41.

First, we prove that $n = 41$ satisfies the condition.

Suppose there exist 15 subsets of X, the union of every 7 of these 15 subsets contains at least 41 elements, and there exists no 3 of the 15 subsets whose intersection is nonempty.

Since every element of X can only belong 2 of the 15 subsets, we can suppose that every element of X belongs to exactly 2 of the 15 subsets. Otherwise we can add a few more elements to some sets of the 15 subsets, and the condition still holds.

By the Pigeonhole Principle, there is a set of the 15 subsets (suppose it is A), such that $|A| \geqslant \left[\dfrac{2\times 56}{15}\right]+1 = 8$. Let the other 14 sets be A_1, A_2, $\cdots$, A_{14}.

The union of every 7 sets of A_1, A_2, $\cdots$, A_{14} contains at least 41 elements. So the total number $y \leqslant 41C_{14}^7$.

We use another way to compute the value of y.

For $a \in X$, if $a \notin A$, then a belongs to exactly 2 sets of A_1, A_2, $\cdots$, A_{14}. So a is counted $C_{14}^7 - C_{12}^7$ times. If $a \in A$, then a belongs to only one set of A_1, A_2, $\cdots$, A_{14}. So a is counted $C_{14}^7 - C_{13}^7$ times. It follows that

$$41C_{14}^7 \leqslant y \leqslant (56 - |A|)(C_{14}^7 - C_{12}^7) + |A|(C_{14}^7 - C_{13}^7)$$

$$= 56(C_{14}^7 - C_{12}^7) - |A|(C_{13}^7 - C_{12}^7)$$

$$\leqslant 56(C_{14}^7 - C_{12}^7) - 8(C_{13}^7 - C_{12}^7),$$

that is, $196 \leqslant 195$, a contradiction.

Next, we prove $n \geqslant 41$.

If $n \leqslant 40$, suppose $X = \{1, 2, \cdots, 56\}$. let

$$A_i = \{i, i+7, i+14, i+21, i+28, i+35, i+42, i+49\},$$

$$i = 1, 2, \cdots, 7,$$

$$B_j = \{j, j+8, j+16, j+24, j+32, j+40, j+48\},$$

$$j = 1, 2, \cdots, 8.$$

It is easy to see that

$$|A_i| = 8 \ (i = 1, 2, \cdots, 7), \ |A_i \cap A_j| = 0 (1 \leqslant i < j \leqslant 7),$$

$$|B_j| = 7 \ (j = 1, 2, \cdots, 8), \ |B_i \cap B_j| = 0 (1 \leqslant i < j \leqslant 8),$$

$$|A_i \cap B_j| = 1 \ (1 \leqslant i \leqslant 7, \ 1 \leqslant j \leqslant 8).$$

For every 3 of the 15 subsets, there are 2 sets both being A_i, or both being B_j. So the intersection of the 3 sets is empty.

But for every 7 of the 15 subsets, for example,

$$A_{i_1}, A_{i_2}, \cdots, A_{i_s}, B_{j_1}, B_{j_2}, \cdots, B_{j_t} \ (s + t = 7),$$

we have

$$|A_{i_1} \cup A_{i_2} \cup \cdots \cup A_{i_s} \cup B_{j_1} \cup B_{j_2} \cup \cdots \cup B_{j_t}|$$

$$= |A_{i_1}| + |A_{i_2}| + \cdots + |A_{i_s}| + |B_{j_1}| + |B_{j_2}| + \cdots + |B_{j_t}| - st$$

$$= 8s + 7t - st = 8s + 7(7 - s) - s(7 - s)$$
$$= (s - 3)^2 + 40 \geqslant 40.$$

So $n \geqslant 41$.

There fore the answer is 41.

China Girls' Mathematical Olympiad

The China Girls' Mathematical Olympiad, organized by the China Mathematical Olympiad Committee, is held in August every year. Participants from Russia, Philippine, Hong Kong, Macau and the mainland are invited to take part in the competition. The competition lasts for 2 days, and there are 4 problems to be completed within 4 hours each day.

2002 (Zhuhai, Guangdong)

The 1st China Girls' Mathematical Olympiad was held on August 15 – 19, 2002 in Zhuhai, Guangdong Province.

The Competition Committee consisted of the following: Pan

Chengbiao, Qiu Zonghu, Li Shenghong, Xiong Bin, Wu Weichao, Qian Zhanwang and Qi Jianxin.

First Day
8:00 – 12:00 August 16, 2002

1 Find all positive integers n such that $20n + 2$ can divide $2\,003n + 2\,002$. (posed by Wu Weichao)

Solution It is easy to see that n is an even number. Let $n = 2m$, then from

$$40m + 2 \mid 2\,003 \times 2m + 2\,002$$

we can get

$$20m + 1 \mid 2\,003m + 1\,001.$$

But

$$2\,003m + 1\,001 = 100(20m + 1) + 3m + 901,$$

so

$$20m + 1 \mid 3m + 901.$$

And when $\dfrac{3m + 901}{20m + 1} = 1, 2, 3, 4$, m is not a positive integer. Therefore,

$$\frac{3m + 901}{20m + 1} \geqslant 5,$$

$$m \leqslant \frac{896}{97} < 10.$$

Hence $m \leqslant 9$. But after checking one by one for $m = 1, 2, 3, \cdots$, 9, we know that $20m + 1 \nmid 3m + 901$ for $m = 1, 2, 3, \cdots, 9$.

Therefore, there is no positive integer n satisfying the condition.

2 $3n$ (n is a positive integer) girl students took part in a summer camp. There were three girl students to be on duty every day. When the summer camp ended, it was found that any two of the $3n$ girl students had just one time to be on duty on the same day.

(1) When $n = 3$, is there any arrangement satisfying the requirement above? Prove your conclusion.

(2) Prove that n is an odd number. (posed by Liu Jiangfeng)

Solution (1) When $n = 3$, there is an arrangement satisfying the requirement. Now the concrete arrangement is given as follows (Denote the nine girl students by $1, 2, \cdots, 9$):

$$(1, 2, 3), (1, 4, 5), (1, 6, 7), (1, 8, 9),$$
$$(2, 4, 6), (2, 7, 8), (2, 5, 9), (3, 4, 8),$$
$$(3, 5, 7), (3, 6, 9), (4, 7, 9), (5, 6, 8).$$

(2) We take arbitrarily one girl student. Since she is on duty just one time with every one of other girl students, and three people are on duty every day. So all other girl students can be paired up two by two. Hence

$$2 \mid 3n - 1,$$

therefore n is an odd number.

Remark Problem (1) is an open-ended problem. It asks whether there is an arrangement satisfying the requirement of the problem. If an arrangements exists, we must construct a concrete example (see the example we have given). If not, we will prove it. (It is usually proved by contradiction.)

🌑 Find all positive integers k such that, for any positive numbers a, b and c satisfying the inequality $k(ab + bc + ca) > 5(a^2 + b^2 + c^2)$, there must exist a triangle with a, b and c as the length of its three sides respectively. (posed by Qian Zhangwang)

Analysis First, we try to determine the range of k. Then by assumption that k is a positive integer, we may get the values for k. Afterward, we check that k can take really these values.

Solution From $(a - b)^2 + (b - c)^2 + (c - a)^2 \geqslant 0$, we get

$$a^2 + b^2 + c^2 \geqslant ab + bc + ca,$$

so $k > 5$. Hence $k \geqslant 6$.

Since there is no triangle with length of its sides to be 1, 1, 2 respectively, by the assumption in the problem, we have

$$k(1 \times 1 + 1 \times 2 + 1 \times 2) \leqslant 5(1^2 + 1^2 + 2^2),$$

that is, $k \leqslant 6$.

We will prove that $k = 6$ satisfies the requirement below. There is no harm in assuming $a \leqslant b \leqslant c$.

Since $\qquad 6(ab + bc + ca) > 5(a^2 + b^2 + c^2),$

so $\qquad 5c^2 - 6(a+b)c + 5a^2 + 5b^2 - 6ab < 0,$

$$\Delta = [6(a+b)]^2 - 4 \cdot 5(5a^2 + 5b^2 - 6ab)$$

$$= 64[ab - (a-b)^2]$$

$$\leqslant 64ab \leqslant 64 \cdot \frac{(a+b)^2}{4}$$

$$= 16(a+b)^2.$$

Thus $\qquad c < \dfrac{6(a+b) + \sqrt{\Delta}}{10} \leqslant \dfrac{6(a+b) + 4(a+b)}{10}$

$$= a + b.$$

Hence, there exists a triangle with a, b, and c being the length of three sides.

Circles O_1 and O_2 interest at two points B and C, and BC is the diameter of circle O_1. Construct a tangent line of circle O_1 at C and interesting circle O_2 at another point A. We join AB to intersect circle O_1 at point E, then join CE and extend it to intersect circle O_2 at point F. Assume H is an arbitrary point on line segment AF. We join HE and extend it to intersect circle O_1 at point G, and then join BG and extend it to intersect the extend line of AC at point D. Prove:

$$\frac{AH}{HF} = \frac{AC}{CD}. \qquad \text{(posed by Xiong Bin)}$$

Analysis To prove $\dfrac{AH}{HF} = \dfrac{AC}{CD}$, we need only to prove $\dfrac{AH}{AF} = \dfrac{AC}{AD}$,

that is,

$$AH \cdot AD = AC \cdot AF.$$

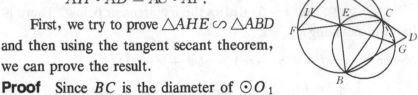

First, we try to prove $\triangle AHE \backsim \triangle ABD$ and then using the tangent secant theorem, we can prove the result.

Proof Since BC is the diameter of $\odot O_1$ and ACD is a tangent line, so $BC \perp AD$, $\angle ACB = 90°$. Hence AB is the diameter of $\odot O_2$.

Also, $\angle BEC = 90°$, that is, $AB \perp FC$, so

$$\angle FAB = \angle CAB.$$

Join CG, then $CG \perp BD$. Hence

$$\angle ADB = \angle BCG = \angle BEG = \angle AEH,$$

therefore $\triangle AHE \backsim \triangle ABD$.

Thus $\dfrac{AH}{AE} = \dfrac{AB}{AD}$,

that is, $AH \cdot AD = AB \cdot AE.$

By the tangent secant theorem, we can get

$$AC^2 = AE \cdot AB,$$

so $AH \cdot AD = AC^2 = AC \cdot AF,$

$$\dfrac{AH}{AF} = \dfrac{AC}{AD}.$$

We can get $\dfrac{AH}{HF} = \dfrac{AC}{CD}.$

Second Day
8:00 – 12:00 August 17, 2002

⑤ Assume P_1, P_2, $\cdots$, $P_n (n \geqslant 2)$ is an arbitrary permutation for

1, 2, $\cdots$, n. Prove that

$$\frac{1}{P_1+P_2}+\frac{1}{P_2+P_3}+\cdots+\frac{1}{P_{n-2}+P_{n-1}}+\frac{1}{P_{n-1}+P_n}>\frac{n-1}{n+2}.$$

(posed by Qiu Zonghu)

Proof By Cauchy's inequality, we can get

$$[(P_1+P_2)+(P_2+P_3)+\cdots+(P_{n-1}+P_n)]\cdot$$

$$\left(\frac{1}{P_1+P_2}+\frac{1}{P_2+P_3}+\cdots+\frac{1}{P_{n-1}+P_n}\right)\geqslant(n-1)^2.$$

Therefore $\qquad \dfrac{1}{P_1+P_2}+\dfrac{1}{P_2+P_3}+\cdots+\dfrac{1}{P_{n-1}+P_n}$

$$\geqslant\frac{(n-1)^2}{2(P_1+P_2+\cdots+P_n)-P_1-P_n}$$

$$=\frac{(n-1)^2}{n(n+1)-P_1-P_n}$$

$$\geqslant\frac{(n-1)^2}{n(n+1)-1-2}$$

$$=\frac{(n-1)^2}{(n-1)(n+2)-1}$$

$$>\frac{(n-1)^2}{(n-1)(n+2)}=\frac{n-1}{n+2}.$$

🔵 Find all pairs of positive integers (x, y) satisfying $x^y=y^{x-y}$.

(posed by Pan Chengbiao)

Solution If $x=1$, then $y=1$. If $y=1$, then $x=1$.

If $x=y$, then $x^y=1$, so $x=y=1$.

We will discuss the circumstances when $x>y\geqslant2$ below. By assumption

$$1<\left(\frac{x}{y}\right)^y=y^{x-2y},$$

so $\qquad\qquad\qquad x>2y$, and $y\mid x$.

Assume $x = ky$ then $k \geqslant 3$, and

$$k^y = y^{(k-2)y}.$$

Therefore $k = y^{k-2}.$

Since $y \geqslant 2$, so $y^{k-2} \geqslant 2^{k-2}$. By mathematical induction or the binomial theorem, it is easy to prove that when $k \geqslant 5$, $2^k > 4k$. Hence k can only be 3 or 4.

When $x = 9$, $y = 3$, we have $k = 3$. When $y = 2$, $x = 8$, we have $k = 4$.

Therefore, all pairs of positive integers to be found are $(1, 1)$, $(9, 3)$ and $(8, 2)$.

An acute triangle ABC has three heights AD, BE and CF respectively. Prove that the perimeter of triangle DEF is not over half of the perimeter of triangle ABC. (posed by Qi Jianxin)

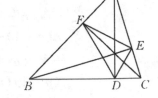

Analysis   To prove

$$DE + EF + DF \leqslant \frac{1}{2}(AB + BC + CA),$$

we need only to prove

$$DE + DF \leqslant BC,$$

$$EF + DF \leqslant AB,$$

$$EF + DE \leqslant AC.$$

Proof   Since $\angle ADB = \angle AEB = 90°$, so four points A, B, D and E are concyclic, and furthermore, AB is the diameter. Hence, by the sine rule, we can get

$$\frac{DE}{\sin \angle DAE} = AB = c,$$

so $DE = c \sin \angle DAE.$

In addition, $\angle DAC + \angle DCA = 90°$, therefore,

$$DE = c\cos C.$$

Similarly, we can get $DF = b\cos B$.

Therefore,
$$DE + DF = c\cos C + b\cos B$$
$$= (2R\sin C)\cos C + (2R\sin B)\cos B$$
$$= R(\sin 2C + \sin 2B)$$
$$= 2R\sin(B+C)\cos(B-C)$$
$$= 2R\sin A\cos(B-C)$$
$$= a\cos(B-C) \leqslant a,$$

where R is the radius of the circumcircle of $\triangle ABC$.

That is,
$$DE + DF \leqslant a.$$

Similarly,
$$DE + EF \leqslant b \text{ and } EF + DF \leqslant c.$$

Therefore,
$$DE + DF + EF \leqslant \frac{1}{2}(a+b+c).$$

Assume that A_1, A_2, $\cdots$, A_8 are eight points taken arbitrarily on a plane. For a directed line l taken arbitrarily on the plane, assume that projections of A_1, A_2, $\cdots$, A_8 on the line are P_1, P_2, $\cdots$, P_8 respectively. If the eight projections are pairwise disjoint, they can be arranged as P_{i_1}, P_{i_2}, $\cdots$, P_{i_8} according to the direction of line l. Thus we get one permutation for 1, 2, $\cdots$, 8, namely, i_1, i_2, $\cdots$, i_8. In the figure, this permutation is $2, 1, 8, 3, 7, 4, 6, 5$. Assume that after these eight points are projected to every directed line on the plane, we get the number of different permutations as $N_8 = N(A_1, A_2, \cdots, A_8)$. Find the maximal value of N_8. (posed by Su Chun)

Solution   (1) For two parallel and

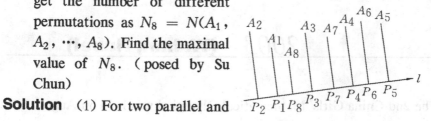

directed lines with the same direction, the order of projections of A_1, A_2, $\cdots$, A_8 must be the same. So, we need only to discuss all directed lines passing through a fixed point O.

(2) If a directed line taken is perpendicular to a line joining two given points, then the projections of these two points must coincide, and a corresponding permutation will not be produced. When the directed lines taken are not perpendicular to any line joining two given points, any two projections of A_1, A_2, $\cdots$, A_8 must not coincide. Hence there is a corresponding permutation.

(3) Suppose that the number of lines through point O and perpendicular to a line joining two given points is k. Then $k \leqslant C_8^2 = 28$. Then there arise $2k$ directed lines placed anticlockwise. Assume that they are in order of l_1, l_2, $\cdots$, l_{2k}. For an arbitrary directed line l (different to l_1, $\cdots$, l_{2k}), there must be two consecutive directed lines l_j and l_{j+1} such that l_j, l, l_{j+1} are placed anticlockwise. It is obvious that for given j, the corresponding permutations obtained from such l must be the same.

(4) For any two directed lines l and l' different from l_1, $\cdots$, l_{2k}, if we cannot find j such that both l_j, l, l_{j+1} and l_j, l', l_{j+1} satisfy (3). Then there must be j so that l', l_j, l are placed anticlockwise. Assume that l_j is perpendicular to the line joining A_{j_1} and A_{j_2}, it is obvious that the orders of the projections of points A_{j_1} and A_{j_2} on the directed lines l and l' must be different, so the corresponding permutations must also be different.

(5) It follows from (3) and (4) that the number of different permutations is $2k$. Note that $k = C_8^2$ is obtainable, so $N_8 = 56$.

2003 (Wuhan, Hubei)

The 2nd China Girls' Mathematical Olympiad was held on August 25 –

29, 2003 in Wuhan, Hubei Province, and was hosted jointly by CMO Committee, Wuhan Education Bureau, Wuhan Iron and Steel Company Limited (WISCO), and the No. 3 Middle School of WISCO.

The Competition Committee consisted of the following: Chen Yonggao, Qian Zhanwang, Li Shenghong, Xiong Bin, Leng Gangsong, Li Weigu, Feng Zuming, and Zhang Zhenjie.

First Day

8:30 – 12:30 August 27, 2003

⬛ Let ABC be a triangle. Points D and E are on sides AB and AC, respectively, and point F is on line segment DE. Let $\dfrac{AD}{AB} = x$, $\dfrac{AE}{AC} = y$, $\dfrac{DF}{DE} = z$. Prove that

(1) $S_{\triangle BDF} = (1 - x)y\, z S_{\triangle ABC}$ and $S_{\triangle CEF} = x(1 - y)(1 - z)S_{\triangle ABC}$;

(2) $\sqrt[3]{S_{\triangle BDF}} + \sqrt[3]{S_{\triangle CEF}} \leqslant \sqrt[3]{S_{\triangle ABC}}$. (posed by Li Weigu)

Solution Connect BE and CD. Then we have

(1) $S_{\triangle BDF} = z S_{\triangle BDE} = z(1 - x)S_{\triangle ABE}$
$\qquad = z(1 - x)y S_{\triangle ABC}$ and

$S_{\triangle CEF} = (1 - z)S_{\triangle CDE}$
$\qquad = (1 - z)(1 - y)S_{\triangle ACD}$
$\qquad = (1 - z)(1 - y)x S_{\triangle ABC}$.

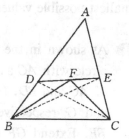

(2) From (1) we get

$$\sqrt[3]{S_{\triangle BDF}} + \sqrt[3]{S_{\triangle CEF}}$$
$$= \left(\sqrt[3]{(1 - x)yz} + \sqrt[3]{x(1 - z)(1 - y)} \right) \sqrt[3]{S_{\triangle ABC}}$$
$$\leqslant \left(\frac{(1 - x) + y + z}{3} + \frac{x + (1 - y) + (1 - z)}{3} \right) \sqrt[3]{S_{\triangle ABC}}$$
$$= \sqrt[3]{S_{\triangle ABC}}.$$

There are 47 students in a classroom with seats arranged in 6 rows ×8 columns, and the seat in the i-th row and j-th column is denoted by (i, j). Now, an adjustment is made for students' seats in the new school term. For a student with the original seat (i, j), if his/her new seat is (m, n), we say that the student is moved by $[a, b] = [i-m, j-n]$ and define the position value of the student as $a+b$. Let S denote the sum of the position values of all the students. Determine the difference between the greatest and smallest possible values of S. (posed by Chen Yonggao)

Solution Add a virtual student A so that every seat is occupied by exactly one student. Denote S' the sum of the position values in this situation. Notice that an exchange of two students occupying the adjacent seats will not change the value of S'. Every student can return to his/her original seat by a finite number of such exchanges of adjacent students. Then $S' \equiv 0$. Since $S' = S + a_A + b_A$, where $a_A + b_A$ is the position value of student A, then we have S is the greatest when student A occupies seat $(1, 1)$, and S is the smallest when A occupies seat $(6, 8)$. So the difference between the greatest and the smallest possible values of S is 14.

As shown in the figure, quadrilateral $ABCD$ is inscribed in a circle with AC as its diameter, $BD \perp AC$, and E the intersection of AC and BD. Extend line segment DA and BA through A to F and G respectively, such that $DG /\!/$ BF. Extend GF to H such that $CH \perp$ GH. Prove that points B, E, F and H lie on one circle. (posed by Liu Jiangfeng)

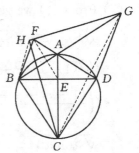

Solution As shown in the figure, connect BH, EF and CG. Since $\triangle BAF \backsim \triangle GAD$, we have

$$\frac{FA}{AB} = \frac{DA}{AG}. \qquad \qquad ①$$

Furthermore, $\triangle ABE \backsim \triangle ACD$, then

$$\frac{AB}{EA} = \frac{AC}{DA}. \qquad \qquad ②$$

Multiplying ① by ②, we get $\frac{FA}{EA} = \frac{AC}{AG}$, then $\triangle FAE \backsim \triangle CAG$

as $\angle FAE = \angle CAG$, and thus $\angle FEA = \angle CGA$.

As is known that $\angle CBG = \angle CHG = 90°$, then points B, C, G and H lie on one circle. So,

$$\angle BHF + \angle BEF = \angle BHC + 90° + \angle BEF$$
$$= \angle BGC + 90° + \angle BEF$$
$$= \angle FEA + 90° + \angle BEF = 180°.$$

That means that points B, E, F and H lie on one circle.

● (1) Prove that there exist five nonnegative real numbers a, b, c, d and e with their sum equal to 1 such that for any arrangement of these numbers around a circle, there are always two neighboring numbers with their product not less than $\frac{1}{9}$.

(2) Prove that for any five nonnegative real numbers with their sum equal to 1, it is always possible to arrange them around a circle such that there are two neighboring numbers with their product not greater than $\frac{1}{9}$. (posed by Qian Zhanwang)

Solution (1) Let $a = b = c = \frac{1}{3}$, $d = e = 0$, it is easy to see that, when arrange them around a circle, we can always get two neighboring numbers of $\frac{1}{3}$, and their product is $\frac{1}{9}$.

(2) For any five nonnegative real numbers a, b, c, d and e

with their sum equal to 1, without loss of generality, we assume that $a \geqslant b \geqslant c \geqslant d \geqslant e \geqslant 0$. Arrange these numbers around a circle in such a way as seen in the figure, we are going to prove that this arrangement meet the condition.

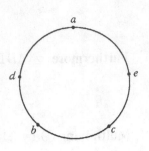

Since $a+b+c+d+e = 1$, we have $a + 3d \leqslant 1$, and

$$a \cdot 3d \leqslant \left(\frac{a+3d}{2}\right)^2 \leqslant \frac{1}{4}, \text{ then } ad \leqslant \frac{1}{12}.$$

Furthermore, $a+b+c \leqslant 1$, then $b+c \leqslant 1-a \leqslant 1-\frac{b+c}{2}$, i.e.
$b+c \leqslant \frac{2}{3}$. So $bc \leqslant \frac{(b+c)^2}{4} \leqslant \frac{1}{9}$. Since $ce \leqslant ae \leqslant ad$ and $bd \leqslant bc$, then any neighboring numbers in this arrangement have their product less than $\frac{1}{9}$.

Second Day
8:30 – 12:30 August 28, 2003

Let $\{a_n\}_1^\infty$ be a sequence of real numbers such that $a_1 = 2$, $a_{n+1} = a_n^2 - a_n + 1$, for $n = 1, 2, \cdots$. Prove that $1 - \dfrac{1}{2\,003^{2\,003}} < \dfrac{1}{a_1} + \dfrac{1}{a_2} + \cdots + \dfrac{1}{a_{2\,003}} < 1$. (posed by Li Shenghong)

Solution We have $a_{n+1} - 1 = a_n(a_n - 1)$, and then $\dfrac{1}{a_{n+1} - 1} = \dfrac{1}{a_n - 1} - \dfrac{1}{a_n}$. So,

$$\frac{1}{a_1} + \frac{1}{a_2} + \cdots + \frac{1}{a_{2\,003}} = \left(\frac{1}{a_1 - 1} - \frac{1}{a_2 - 1}\right) + \left(\frac{1}{a_2 - 1} - \frac{1}{a_3 - 1}\right)$$

$$+ \cdots + \left(\frac{1}{a_{2\,003} - 1} - \frac{1}{a_{2\,004} - 1}\right)$$

$$= \frac{1}{a_1 - 1} - \frac{1}{a_{2\,004} - 1} = 1 - \frac{1}{a_{2\,004} - 1}.$$

It is easy to see that $\{a_n\}$ is strictly monotonic increasing, so $a_{2\,004} > 1$, and that means

$$\frac{1}{a_1} + \frac{1}{a_2} + \cdots + \frac{1}{a_{2\,003}} < 1.$$

In order to prove the left inequality, we only need to prove that $a_{2\,004} - 1 > 2\,003^{2\,003}$.

By induction, we have $a_{n+1} = a_n a_{n-1} \cdot \cdots \cdot a_1 + 1$ and $a_n a_{n-1} \cdot \cdots \cdot a_1 > n^n \, (n \geqslant 1)$.

This completes the proof.

⬤ Let $n \geqslant 2$ be an integer. Find the largest real number λ such that the inequality

$$a_n^2 \geqslant \lambda(a_1 + a_2 + \cdots + a_{n-1}) + 2a_n$$

holds for any positive integers a_1, a_2, $\cdots$, a_n satisfying $a_1 < a_2 < \cdots < a_n$. (posed by Leng Gangsong)

Solution For $a_i = i$, $i = 1, 2, \cdots, n$, we have $\lambda \leqslant (n-2) \div \frac{n-1}{2} = \frac{2n-4}{n-1}$.

Now, we are going to prove that $a_n^2 \geqslant \frac{2n-4}{n-1}(a_1 + a_2 + \cdots + a_{n-1}) + 2a_n$ for any positive integers a_1, a_2, $\cdots$, a_n satisfying $a_1 < a_2 < \cdots < a_n$.

Since $a_k \leqslant a_n - (n-k)$, $k = 1, 2, \cdots, n-1$, $a_n \geqslant n$, then we have

$$\frac{2n-4}{n-1} \sum_{i=1}^{n-1} a_i \leqslant \frac{2n-4}{n-1}\left((n-1)a_n - \frac{n(n-1)}{2}\right)$$

$$= (2n-4)a_n - n(n-2)$$

$$= (n-2)(2a_n - n) \leqslant (a_n - 2)a_n.$$

That is, $a_n^2 \geqslant \dfrac{2n-4}{n-1}(a_1 + a_2 + \cdots + a_{n-1}) + 2a_n$. So the largest

value of λ is $\dfrac{2n-4}{n-1}$.

7 Let the sides of a scalene triangle $\triangle ABC$ be $AB = c$, $BC = a$, $CA = b$, D, E, F be points on BC, CA, AB, such that AD, BE, CF are angle bisectors of the triangle, respectively. Assume that $DE = DF$. Prove that

(1) $\dfrac{a}{b+c} = \dfrac{b}{c+a} + \dfrac{c}{a+b}$;

(2) $\angle BAC > 90°$.

(posed by Xiong Bin)

Solution Using the sine rule, we have that

$$\frac{\sin \angle AFD}{\sin \angle FAD} = \frac{AD}{FD} = \frac{AD}{ED} = \frac{\sin \angle AED}{\sin \angle FAD},$$

then $\sin \angle AFD = \sin \angle AED$. So, either $\angle AFD = \angle AED$ or $\angle AFD + \angle AED = 180°$.

If $\angle AFD = \angle AED$, then $\triangle ADF \cong \triangle ADE$, and we get $AF = AE$. Then $\triangle AIF \cong \triangle AIE$, and $\angle AFI = \angle AEI$. So $\triangle AFC \cong \triangle AEB$, then $AC = AB$. It contradicts the condition given. So $\angle AFD + \angle AED = 180°$, and points A, F, D and E lie on one circle.

Then $\angle DEC = \angle DFA > \angle ABC$. Extend CA through A to point P such that

$$\angle DPC = \angle B, \text{ then } PC = PE + CE. \qquad \textcircled{1}$$

Since $\angle BFD = \angle PED$ and $FD = ED$, we have that $\triangle BFD \cong \triangle PED$, then $PE = BF = \dfrac{ac}{a+b}$. Furthermore, $\triangle PCD \backsim \triangle BCA$, then $\dfrac{PC}{BC} = \dfrac{CD}{CA}$. So

$$PC = a \cdot \frac{ba}{b+c} \cdot \frac{1}{b} = \frac{a^2}{b+c}. \qquad \textcircled{2}$$

From ① and ② we get $\dfrac{a^2}{b+c} = \dfrac{ac}{a+b} + \dfrac{ab}{c+a}$, then $\dfrac{a}{b+c} = \dfrac{b}{c+a} + \dfrac{c}{a+b}$.

This completes the proof of (1).

As to the proof of (2), we have from (1) that

$$a(a+b)(a+c) = b(b+a)(b+c) + c(c+a)(c+b),$$

$$a^2(a+b+c) = b^2(a+b+c) + c^2(a+b+c) + abc$$

$$> b^2(a+b+c) + c^2(a+b+c).$$

Then $a^2 > b^2 + c^2$, and that means $\angle BAC > 90°$.

🔵 Let n be a positive integer, and S_n be the set of all positive integer divisors of n (including 1 and itself). Prove that at most half of the elements in S_n have their last digits equal to 3. (posed by Feng Zuming)

Solution We are going to consider the following three cases.

(1) If $5 \mid n$, let $d_1, d_2, \cdots, d_m$ be the elements in S_n with their last digits equal to 3, then $5d_1, 5d_2, \cdots, 5d_m$ are elements in S_n with their last digits equal to 5. So $m \leqslant \dfrac{1}{2} |S_n|$. The statement is true in this case.

(2) If $5 \nmid n$ and the last digit of every prime divisor of n is either 1 or 9, the last digit of any element in S_n is either 1 or 9. The statement is also true in this case.

(3) If $5 \nmid n$ and there exists a prime divisor p in S_n such that the last digit of p is either 3 or 7. Let $n = p^r q$, where q and r are positive integers and p is prime to q, and let $S_q = \{a_1, a_2, \cdots, a_k\}$ be the set of all positive integer divisors of q. Then the elements in S_n can be written in the following way:

$$a_1, a_1 p, a_1 p^2, \cdots, a_1 p^r,$$

$$a_2, a_2 p, a_2 p^2, \cdots, a_2 p^r,$$

$$\cdots\cdots\cdots\cdots\cdots\cdots\cdots\cdots$$

$$a_k, a_k p, a_k p^2, \cdots, a_k p^r.$$

For any $d_i = a_s p^l \in S_n$, we choose $e_i = \begin{cases} a_s p^{l+1} & l < r, \\ a_s p^{l-1} & l = r, \end{cases}$ then $e_i \in$ S_n and we call e_i the partner of d_i. If the last digit of d_i is 3, then that of its partner e_i is not, since that of p is either 3 or 7. If d_i and d_j in S_n are different, and their last digits are both 3, then their partners e_i and e_j are also different. Otherwise, suppose $e_i = e_j = a_s p^l$, we may assume that $\{d_i, d_i\} = \{a_s p^{l-1}, a_s p^{l+1}\}$, then $d_i = d_j p^2$. As the last digit of p is 3 or 7, then that of p^2 is always 9, and that means the last digits of d_i and d_j cannot be the same. It leads to a contradiction.

We then see that every $d_i \in S_n$ with its last digit equal to 3 has a partner $e_i \in S_n$ with its last digit not equal to 3, and different d_i has different partner. That means that at most half of the elements in S_n have their last digits equal to 3. This completes the proof.

2004   (Nanchang, Jiangxi)

The 3rd China Girls' Mathematical Olympiad was held on August 10 – 15, 2004 in Nanchang, Jiangxi Province, China, and was hosted jointly by CMO Committee, Jiangxi Mathematical Society and Nanchang No. 2 middle school.

The Competition Committee consisted of the following: Tao Pingsheng, Su Chun, Chen Yonggao, Xiong Bin, Li Shenghong, Li Weigu, Shi qiyan, Yuan Hanhui and Feng Zuming.

First Day
8:00 – 12:00 August 12, 2004

⬤ We say a positive integer n is "good" if there is a permutation

$(a_1, a_2, \cdots, a_n)$ of $1, 2, \cdots, n$ such that $a_k + k$ is a perfect square for all $1 \leqslant k \leqslant n$. Determine all the good numbers in the set $\{11, 13, 15, 17, 19\}$. (posed by Su Chun)

Solution The good numbers are 13, 15, 17 and 19. However 11 is not.

Note that for $1 \leqslant k \leqslant 11$, $4 + k$ is a perfect square if and only if $k = 5$. Likewise, $11 + k$ is a perfect square if and only if $k = 5$. Hence 11 is not good.

Note that 13 is good because

k:	1	2	3	4	5	6	7	8	9	10	11	12	13,
a_k:	8	2	13	12	11	10	9	1	7	6	5	4	3.

Similarly, 15, 17 and 19 are good because

k:	1	2	3	4	5	6	7	8	9	10	11	12	13	14	15,			
a_k:	15	14	13	12	11	10	9	8	7	6	5	4	3	2	1;			
k:	1	2	3	4	5	6	7	8	9	10	11	12	13	14	15	16	17,	
a_k:	3	7	6	5	4	10	2	17	16	15	14	13	12	11	1	9	8;	
k:	1	2	3	4	5	6	7	8	9	10	11	12	13	14	15	16	17	18	19,
a_k:	8	7	6	5	4	3	2	1	16	15	14	13	12	11	10	9	19	18	17.

🔷 Let a, b and c be positive real numbers. Determine the minimum value of

$$\frac{a+3c}{a+2b+c} + \frac{4b}{a+b+2c} - \frac{8c}{a+b+3c}. \quad \text{(posed by Li Shenghong)}$$

Solution The answer is $12\sqrt{2} - 17$. Set

$$\begin{cases} x = a + 2b + c, \\ y = a + b + 2c, \\ z = a + b + 3c. \end{cases}$$

It is easy to see that $z - y = c$ and $x - y = b - c$, giving $x - y = b - (z - y)$, or $b = x + z - 2y$. We note that $a + 3c = 2y - x$. By the AM-GM Inequality, it follows that

$$\frac{a+3c}{a+2b+c}+\frac{4b}{a+b+2c}-\frac{8c}{a+b+3c}$$

$$=\frac{2y-x}{x}+\frac{4(x+z-2y)}{y}-\frac{8(z-y)}{z}$$

$$=-17+2\frac{y}{x}+4\frac{x}{y}+4\frac{z}{y}+8\frac{y}{z}$$

$$\geqslant-17+2\sqrt{8}+2\sqrt{32}=-17+12\sqrt{2}.$$

The equality holds if and only if $\frac{2y}{x}=\frac{4x}{y}$ and $\frac{4z}{y}=\frac{8y}{z}$, or $4x^2=2y^2=z^2$. Hence the equality holds if and only if

$$\begin{cases}a+b+2c=\sqrt{2}(a+2b+c),\\a+b+3c=2(a+2b+c).\end{cases}$$

Solving the above system of equations for b and c in terms of a gives

$$\begin{cases}b=(1+\sqrt{2})a,\\c=(4+3\sqrt{2})a.\end{cases}$$

We conclude that

$$\frac{a+3c}{a+2b+c}+\frac{4b}{a+b+2c}-\frac{8c}{a+b+3c}$$

has minimum value $12\sqrt{2}-17$ if and only if $(a,b,c)=(a,(1+\sqrt{2})a,(4+3\sqrt{2})a)$.

Let ABC be an obtuse triangle inscribed in a circle of radius 1. Prove that triangle ABC can be covered by an isosceles right-angled triangle with hypotenuse $\sqrt{2}+1$. (posed by Leng Gangsong)

Solution Without loss of generality, we may assume that $\angle C>90°$. Since $\angle A+\angle B<90°$, we may assume without loss of generality that $\angle A < 45°$.

We can then construct a semicircle ω with AB as its diameter such that point C lies inside ω. Let O be the center of ω. Then $AO = BO$. Construct rays AT and OE such that $\angle BAT = \angle BOE = 45°$ with E lying on ω. Let l be the line tangent to ω at E, and let l meet rays AT and AB at F and D respectively. It is not difficult to see that triangle AFD is an isosceles right-angled triangle, with $\angle F = 90°$, that covers triangle ABC. It suffices to show that $AD < \sqrt{2} + 1$. Note that

$$AD = AO + OD = AO + \sqrt{2}OE = (\sqrt{2} + 1)AO.$$

Applying the sine rule to triangle ABC gives $AB = 2\sin C < 2$, or $AO < 1$. It follows that $AD < \sqrt{2} + 1$, as desired.

> A deck of 32 cards has 2 different jokers each of which is numbered 0. There are 10 red cards numbered 1 through 10 and similarly for blue and green cards. One chooses a number of cards from the deck. If a card in hand is numbered k, then the value of the card is 2^k, and the value of the hand is the sum of the values of cards in hand. Determine the number of hands having the value 2004. (posed by Tao Pingsheng)

Remark The answer of the problem is $1\,003^2 = 1\,006\,009$. We provide two approaches as follows.

Solution I   For each hand h having the value 2004, let j_h, r_h, b_h and g_h denote the total values of jokers, red, blue, and green cards in hand. Then (j_h, r_h, b_h, g_h) is an ordered quadruples of nonnegative integers such that

$$j_h + r_h + b_h + g_h = 2\,004, \qquad (*)$$

where $0 \leqslant j_h \leqslant 2$ and r_h, b_h and g_h are even. Note that $2\,004 < 2\,048 = 2^{11}$, each even nonnegative integer e less than $2\,004$ can be written uniquely in the form of

$$a_1 \cdot 2^1 + a_2 \cdot 2^2 + \cdots + a_{10} \cdot 2^{10},$$

where $a_i = 0$ or 1 for $1 \leqslant i \leqslant 10$. (This is basically the binary representation of e.) Hence each ordered quadruple of nonnegative integers (j_h, r_h, b_h, g_h) satisfying equation (*) can be mapped one-to-one and onto a hand having the value 2 004. It suffices to count the number of ordered quadruples of nonnegative integers in equation (*) that satisfy the conditions. Since r_h, b_h and g_h are even, j_h is also even. Set $j_h = 2j$, $r_h = 2r$, $b_h = 2b$, and $g_h = 2g$. It suffices to count the number of ordered quadruples (j, r, b, g) of nonnegative integers such that

$$j + r + b + g = 1\,002, \qquad (**)$$

with $j = 0$ or 1. We consider two cases.

In the first case, we assume that $j = 0$. Then $r + b + g = 1\,002$ has $\binom{1\,004}{2} = \dfrac{1\,004 \cdot 1\,003}{2}$ ordered triples (r, b, g) of nonnegative integers such that $r + b + g = 1\,002$, that is, there are $\dfrac{1\,004 \cdot 1\,003}{2}$ ordered quadruples $(0, r, b, g)$ of nonnegative integers satisfying equation (**).

In the second case, we assume that $j = 2$. Then $r + b + g = 1\,001$ has $\binom{1\,003}{2} = \dfrac{1\,003 \cdot 1\,002}{2}$ ordered triples (r, b, g) of nonnegative integers such that $r + b = g = 1\,001$, that is, there are $\dfrac{1\,003 \cdot 1\,002}{2}$ ordered quadruples $(1, r, b, g)$ of nonnegative integers satisfying equation (**). Therefore, the answer for the problem is

$$\frac{1\,004 \cdot 1\,003}{2} + \frac{1\,003 \cdot 1\,002}{2} = 1\,003^2.$$

Solution II For a particular hand, a card in hand numbered k contributes either 0 or 2^k points to the value of the hand. We consider the following generating function

$$f(x) = (1+x^{2^0})^2[(1+x^{2^1})(1+x^{2^2})\cdots(1+x^{2^{10}})]^3.$$

That is, if a_n denotes the number of hands having a value n, then $a_n = [x_n]f(x)$, where $[x_n]h(x)$ denotes the coefficients of x^n in the expansion of the polynomial $h(x)$. We compute $a_{2\,004} = [x_{2\,004}]f(x)$. Note that

$$f(x) = \frac{[(1+x^{2^0})(1+x^{2^1})\cdots(1+x^{2^{10}})]^3}{1+x}$$

$$= \frac{[(1-x^{2^0})(1+x^{2^0})(1+x^{2^1})\cdots(1+x^{2^{10}})]^3}{(1+x)(1-x)^3}$$

$$= \frac{(1-x^{2^{11}})^3}{(1+x)(1-x)^3}.$$

Since $2\,004 < 2^{11}$, we have $[x_{2\,004}]f(x) = [x_{2\,004}]g(x)$, where

$$g(x) = \frac{1}{(1+x)(1-x)^3}.$$

Note that

$$g(x) = \frac{1}{(1+x)(1-x)^3} = \frac{1}{(1-x^2)(1-x)^2}$$

$$= (1+x^2+x^4+x^6+\cdots)(1+2x+3x^2+4x^3+\cdots).$$

It follows that for any positive integer k,

$$[x_{2k}]g(x) = 1+3+5+\cdots+(2k+1) = (k+1)^2.$$

In particular, $a_{2\,004} = [x_{2\,004}]g(x) = 1\,003^2$.

Second Day
8:00 – 12:00 August 13, 2004

5 Determine the maximum value of constant λ such that

$$u+v+w \geqslant \lambda,$$

where u, v and w are positive real numbers with $u\sqrt{vw} + v\sqrt{wu} + w\sqrt{uv} \geqslant 1$. (posed by Chen Yonggao)

Remark The maximum value of λ is $\sqrt{3}$. It is not difficult to see that for $u = v = w = \dfrac{\sqrt{3}}{3}$, $u\sqrt{vw} + v\sqrt{wu} + w\sqrt{uv} = 1$ and $u + v + w = \sqrt{3}$.

Hence the maximum value of λ is less than or equal to $\sqrt{3}$. It suffices to show that $u + v + w \geqslant \sqrt{3}$.

Solution I By the AM-GM inequality and the given condition, we have

$$u \cdot \frac{v+w}{2} + v \cdot \frac{w+u}{2} + w \cdot \frac{u+v}{2}$$

$$\geqslant u\sqrt{vw} + v\sqrt{wu} + w\sqrt{uv} \geqslant 1,$$

or

$$uv + vw + wu \geqslant 1.$$

Since

$$(u-v)^2 + (v-w)^2 + (w-u)^2 \geqslant 0,$$

$$u^2 + v^2 + w^2 \geqslant uv + vw + wu \geqslant 1,$$

it follows that

$$(u+v+w)^2 = u^2 + v^2 + w^2 + 2(uv + vw + wu) \geqslant 3,$$

or $u + v + w \geqslant \sqrt{3}$. The equality holds if and only if $u = v = w = \dfrac{\sqrt{3}}{3}$.

Solution II By the AM-GM inequality and then by Cauchy's Inequality, we have

$$\frac{(u+v+w)^4}{9} = \left(\frac{u+v+w}{3}\right)^3 \cdot 3(u+v+w)$$

$$\geqslant 3uvw(u+v+w)$$

$$= (uvw + vwu + wuv)(u+v+w)$$

$$\geqslant (u\sqrt{vw} + v\sqrt{wu} + w\sqrt{uv})^2 = 1,$$

which implies that $u+v+w \geqslant \sqrt{3}$. The equality hold if and only if $u = v = w = \dfrac{\sqrt{3}}{3}$.

Solution III Set $x = \sqrt{uv}$, $y = \sqrt{vw}$, and $z = \sqrt{wu}$. Then the given condition reads

$$xy + yz + zx = v\sqrt{wu} + w\sqrt{uv} + u\sqrt{vw} \geqslant 1.$$

Note that $u = \dfrac{zx}{y}$, $v = \dfrac{xy}{z}$, and $w = \dfrac{yz}{x}$. By the AM-GM inequality, we have

$$2(u+v+w) = \left(\frac{zx}{y} + \frac{xy}{z} \right) + \left(\frac{xy}{z} + \frac{yz}{x} \right) + \left(\frac{yz}{x} + \frac{zx}{y} \right)$$

$$\geqslant 2x + 2y + 2z,$$

or $u + v + w \geqslant x + y + z$. As shown in Solution I, we have $(x + y + z)^2 \geqslant 3(xy + yz + zx) \geqslant 3$. Hence $(u+v+w)^2 \geqslant (x+y+z)^2 \geqslant 3$, or $u+v+w \geqslant \sqrt{3}$, and the equality holds if and only if $x = y = z$, that is, $u = v = w = \dfrac{\sqrt{3}}{3}$.

⬤ Given an acute triangle ABC with O as its circumcenter. Line AO and side BC meet at D. Points E and F are on sides AB and AC respectively, such that points A, E, D and F are on a circle. Prove that the length of the projection of line segment EF on side BC does not depend on the positions of E and F. (posed by Xiong Bin)

Solution Let M and N be the feet of the perpendiculars from D to lines AB and AC respectively. Let E_0, F_0, M_0 and N_0 be the feet of the

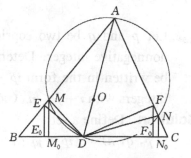

perpendiculars from E, F, M and N to side BC respectively. It suffices to show that $E_0F_0 = M_0N_0$. Without loss of generality, we may assume that E lies on line segment BM. Then $\angle AED < 90°$. Since A, E, D, F are concyclic, $\angle DFC = \angle AED < 90°$, and so F lies on line segment AN. It follows that we need only to consider the figure above. It suffices to show that $E_0M_0 = F_0N_0$. Note that $E_0M_0 = EM \cos \angle B$ and $F_0N_0 = FN\cos \angle C$. We need only to show that $EM\cos \angle B = FN\cos \angle C$, or

$$\frac{EM}{FN} = \frac{\cos\angle C}{\cos\angle B}. \tag{1}$$

Since $\angle MED = \angle AED = \angle DFC = \angle NFD$, the right-angled triangles EMO and FNO are similar, which implies that

$$\frac{EM}{FN} = \frac{MD}{ND}. \tag{2}$$

In the right-angled triangles AMO and ANO, $MD = AD \sin\angle DAM = AD\sin \angle OAB$ and $ND = AD \sin \angle DAN = AD \sin\angle OAC$. Substitute these equations into equation (2) and we have

$$\frac{EM}{FN} = \frac{\sin\angle OAB}{\sin\angle OAC}. \tag{3}$$

Since O is the circumcenter of triangle ABC, $\angle AOB = 2\angle C$ and $\angle OAB = \angle OBA = 90° - \angle C$, and so $\sin\angle OAB = \sin(90° - \angle C) = \cos \angle C$. Likewise, $\sin \angle OAC = \cos \angle B$. Substitute the last two equations into equation (3) and we get the desired equation (1).

🔴 Let p and q be two coprime positive integers, and let n be a nonnegative integer. Determine the number of integers that can be written in the form $ip + jq$, where i and j are nonnegative integers with $i + j \leqslant n$. (posed by Li Weigu)

Solution Define a set

$S(p, q, n) = \{ip + jq \mid i \text{ and } j \text{ are nonnegative integers with } i +$

$j \leqslant n$.

Let $s_n = | S(p, q, n) |$, where $| X |$ denotes the number of elements in set X. The answer of the problem is

$$s_n = \begin{cases} \dfrac{(n+1)(n+2)}{2}, & \text{if } n < r, \\[2mm] \dfrac{r(2n-r+3)}{2}, & \text{if } n \geqslant r, \end{cases} \qquad (*)$$

where $r = \max\{p, q\}$.

Now we establish the equation $(*)$. Without loss of generality, we assume that $r = p > q$. It is easy to see that $s_0 = | S(p, q, 0) | = | \{0\} | = 1$ satisfying equation $(*)$. Note that

$$S(p, q, n) \backslash S(p, q, n-1) \subseteq \{ip + (n-i)q \mid i = 0, 1, \cdots, n\}.$$

Note also that

$$ip + (n-i)q = (i+q)p + (n-p-i)q,$$

with

$$(i+q) + (n-p-i) = n+q-p \leqslant n-1.$$

Hence number $ip + (n-i)q$ belongs to both sets $S(p, q, n)$ and $S(p, q, n-1)$ if and only if $n-p-i \geqslant 0$, or $i \leqslant n-p$. Therefore,

$$S(p, q, n) \backslash S(p, q, n-1)$$

$$= \begin{cases} \{ip + (n-i)q \mid i = n-p+1, n-p+2, \cdots, n\}, & \text{if } n \geqslant p, \\ \{ip + (n-i)q \mid i = 0, 1, \cdots, n\}, & \text{if } n < p. \end{cases}$$

It implies that

$$s_n - s_{n-1} = \begin{cases} p, & \text{if } n \geqslant p, \\ n+1, & \text{if } n < p. \end{cases}$$

If $n < p$, we conclude that

$$s_n = s_0 + (s_1 - s_0) + \cdots + (s_n - s_{n-1})$$

$$=1+2+\cdots+(n+1) = \frac{(n+1)(n+2)}{2}.$$

In particular, $s_{p-1} = \frac{p(p+1)}{2}$. If $n \geqslant p$, we conclude that

$$s_n = s_{p-1} + (s_p - s_{p-1}) + \cdots + (s_n - s_{n-1})$$

$$= \frac{p(p+1)}{2} + (n-p+1)p = \frac{p(2n-p+3)}{2}.$$

Remark   If $p = q$, then

$$S(p, q, n) = S(p, p, n) = \{(i+j)p \mid i \text{ and } j \text{ are nonnegative}$$

integers with $i + j \leqslant n\}$.

It is easy to see that $|S(p, p, n)|$ is equal to the number of ordered triples of nonnegative integers (i, j, k) such that $i + j + k = n$, which implies that $|S(p, p, n)| = \dfrac{(n+1)(n+2)}{2}$.

When the unit squares at the four corners are removed from a three by three square, the resulting shape is called a cross. What is the maximum number of non-overlapping crosses placed within the boundary of a 10×11 chessboard? (Each cross covers exactly five unit squares on the board.) (posed by Feng Zuming)

Solution   The answer is 15.

We first show that it is impossible to place 16 crosses within the boundary of a 10×11 chessboard. We approach indirectly by assuming that we could place 16 crosses.

The centers of the crosses (denoted by $*$) must lie in the 8×9 subboard in the middle. We tile this central board by three 8×3 boards, and label these three boards (a), (b) and (c), from left to right. We consider the number of centers placed in the three boards.

It is easy to see that in a 2×3 board or a 3×3 board, we can place at most two centers.

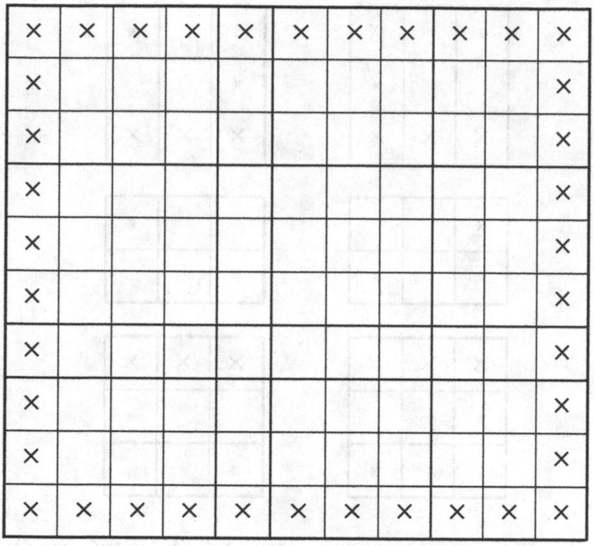

Note that we can place at most two centers on each 3×3 subboard:

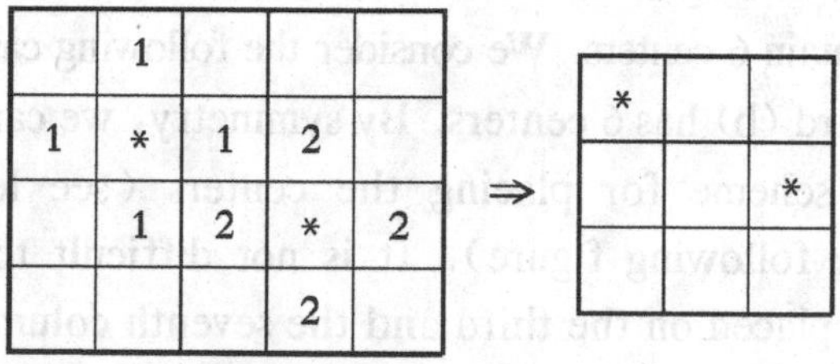

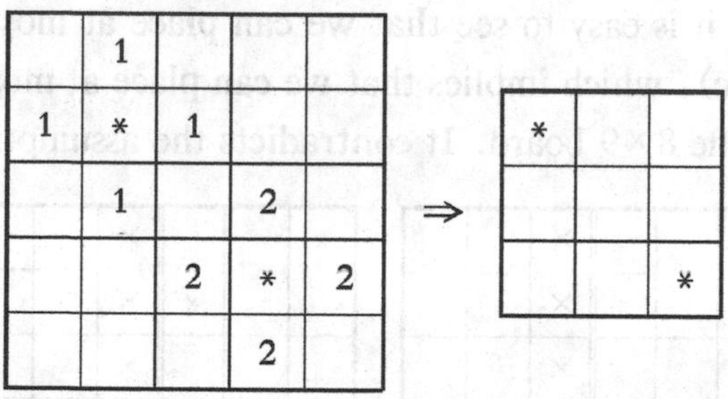

We can tile a 8×3 board one 2×3 board sandwiched by two 3×3 boards. Hence we can place at most 6 centers on a 8×3 board, with each two centers placed on each subboard. Since there are two centers placed in the middle 2×3 subboard, no centers can be places in the third and sixth row of the 8×3 board. We can only have the following two symmetric distributions.

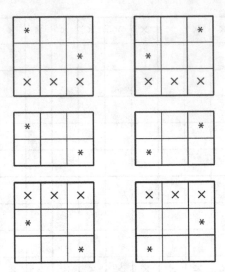

Therefore, there are 16 centers placed in board (a), (b) and (c), with each board holding at most 6 centers. Hence at least one of the three boards must contain 6 centers. We consider the following cases.

Case I Board (b) has 6 centers. By symmetry, we can assume that the following scheme for placing the centers (see left-hand side diagram in the following figure). It is not difficult to see that no centers can be placed on the third and the seventh columns of the 8×9 board. Then it is easy to see that we can place at most 4 centers in board (a) or (c), which implies that we can place at most $4+6+4=14$ centers on the 8×9 board. It contradicts the assumption.

Case II Both boards (a) and (c) have six centers. By symmetry, we discuss with the right-hand side diagram in the following figure. In this case there is no centers can be placed in the forth and the sixth columns of the 8×9 board, which implies that board (b) can hold at most 3 centers, and so the 8×9 board can hold at most $6 + 3 + 6 = 15$ centers, which is again a contradiction.

Case III Exactly one of the boards (a) and (c) has six centers. In this case we assume that (a) contains 6 centers and board (c) contains at most 5 centers. Then no center can be placed in the fourth column of the 8×9 board. It follows that board (b) contains at most 4 centers. Hence there are at most $6 + 4 + 5 = 15$ centers on the 8×9 board, which is again a contradiction.

Combining the above argument, we conclude that it is impossible to place 16 centers on a 8×9 board.

We complete our solution by providing two different ways to place 15 centers on the board.

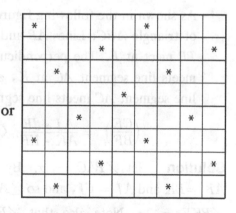

Remark The example shown on the right-hand side comes from the fact that the crosses can tile the whole plane without any gaps. The example shown on the left-hand side shows that it is possible to have holes (or gaps) between the crosses for this particular problem, which makes the other approaches for this problem difficult.

2005 (Changchun, Jilin)

The 4th China Girls' Mathematical Olympiad was held on August 10 – 15, 2005 in Changchun, Jilin Province, China, and was hosted by CMO Committee.

The Competition Committee consisted of the following: Wang Jie (the president), Zhu Huawei, Chen Yonggao, Su Chun, Xiong Bin, Feng Zuming, Zhang Tongjun, Feng Yuefeng, Ye Zhonghao and Yuan Hanhui.

First Day
8:00 – 12:00 August 12, 2005

As shown in the following figure, point P lies on the circumcicle of triangle ABC. Lines AB and CP meet at E, and lines AC and BP meet at F. The perpendicular bisector of line segment AB meets line segment AC at K, and the perpendicular bisector of line segment AC meets line segment AB at J. Prove that

$$\left(\frac{CE}{BF}\right)^2 = \frac{AJ \cdot JE}{AK \cdot KF}. \quad \text{(posed by Ye Zhonghao)}$$

Solution   Set $\angle BAC = x$. By the given condition, we have $AK = BK$ and $AJ = CJ$, and so $\angle ABK = \angle ACJ = x$ and $\angle EJC = \angle BKF = 2x$. Note also that $\angle ECJ = \angle ACP - \angle ACJ = \angle ACP - x$. In triangle ABF, we have $\angle AFB = 180° - \angle ABP - \angle ABF = 180° - \angle ABP - x$. Since A, B, P, C are concyclic, we have $\angle ACP = 180° - \angle ABP$. Combining with the above

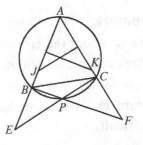

equation, we conclude that $\angle BFK = \angle BFA = \angle ACP - x = \angle ECJ$. Thus, in triangles CEJ and FBK, we have $\angle CJE = \angle BKF$ and $\angle ECJ = \angle BFK$. So the two triangles are similar. It follows that

$$\frac{CE}{FB} = \frac{CJ}{FK} = \frac{EJ}{BK}.$$

Consequently, we have

$$\left(\frac{CE}{BF}\right)^2 = \frac{AJ \cdot JE}{AK \cdot KF}$$

as desired.

⬤ Find all ordered triples (x, y, z) of real numbers such that

$$\begin{cases} 5\left(x+\dfrac{1}{x}\right)= 12\left(y+\dfrac{1}{y}\right)= 13\left(z+\dfrac{1}{z}\right), \\ xy + yz + zx = 1. \end{cases}$$ (posed by Zhu Huawei)

Remark The solutions are $\left(\dfrac{1}{5}, \dfrac{2}{3}, 1\right)$ and $\left(-\dfrac{1}{5}, -\dfrac{2}{3}, -1\right)$.

Assume that (x, y, z) is a solution of the given system. Clearly, $xyz \neq 0$. If $x > 0$, then by the first given equation, we have $y > 0$ and $z > 0$. If $x < 0$, then it is clear that $y < 0$ and $z < 0$, and so $(-x, -y, -z)$ is a solution of the given system.

Solution There are angles A, B, and C in the interval $(0°, 180°)$ such that

$$x = \tan\frac{A}{2}, \ y = \tan\frac{B}{2}, \ z = \tan\frac{C}{2}.$$

By the addition and subtraction formulas, the second equation in the given system becomes

$$1 = \tan\frac{A}{2} \tan\frac{B}{2} + \tan\frac{B}{2} \tan\frac{C}{2} + \tan\frac{C}{2} \tan\frac{A}{2}$$

$$= \tan\frac{A}{2} \tan\frac{B}{2} + \tan\frac{C}{2} \tan\frac{A+B}{2}\left(1 - \tan\frac{A}{2} \tan\frac{B}{2}\right),$$

implying that

$$\left(1 - \tan\frac{C}{2}\tan\frac{A+B}{2}\right)\left(1 - \tan\frac{A}{2}\tan\frac{B}{2}\right) = 0.$$

Note that $\tan\dfrac{A}{2}\tan\dfrac{B}{2} = xy \neq 1$, otherwise $z(x+y) = 0$, implying

$z = 0$, which is impossible. Therefore, $\tan\dfrac{C}{2}\tan\dfrac{A+B}{2} = 1$, or

$\dfrac{A+B}{2} + \dfrac{C}{2} = 90°$. In other words, A, B, C are the angles of a

triangle. Let ABC denote that triangle.

We rewrite the first equation in the given system as

$$\frac{x}{5(x^2+1)} = \frac{y}{12(y^2+1)} = \frac{z}{13(z^2+1)}.$$

Note that, by the double-angle formula, we have

$$\frac{x}{x^2+1} = \frac{\tan\dfrac{A}{2}}{\tan^2\dfrac{A}{2}+1} = \frac{\tan\dfrac{A}{2}}{\sec^2\dfrac{A}{2}} = \sin\frac{A}{2}\cos\frac{A}{2} = \frac{\sin A}{2};$$

and analogously for the expressions of y and z. We conclude that

$$\frac{\sin A}{5} = \frac{\sin B}{12} = \frac{\sin C}{13}.$$

By the sine rule, the sides of triangle ABC are in ratio $5 : 12 : 13$

with $\sin A = \dfrac{5}{13}$, $\sin B = \dfrac{12}{13}$ and $\sin C = 1$. Hence $\dfrac{x}{x^2+1} = \dfrac{5}{26}$, or

$5x^2 - 26x + 5 = 0$, implying that $x = 5$ or $x = \dfrac{1}{5}$. Likewise, we have

$y = \dfrac{3}{2}$ or $y = \dfrac{2}{3}$, and $z = 1$. Substituting $z = 1$ into the second equation

in the given system leads to $xy + x + y = 1$, implying that $(x, y, z) = \left(\dfrac{1}{5}, \dfrac{2}{3}, 1\right)$ is the only solution with $x > 0$. Hence $\left(\dfrac{1}{5}, \dfrac{2}{3}, 1\right)$ and

$\left(-\dfrac{1}{5}, -\dfrac{2}{3}, 1\right)$ are the solutions of the problem.

4 Determine if there exists a convex polyhedron such that

(1) it has 12 edges, 6 faces and 8 vertices;

(2) it has 4 faces with each pair of them sharing a common edge of the polyhedron. (posed by Su Chun)

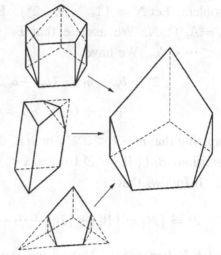

Solution The answer is yes, as shown in the figure.

5 Determine all positive real numbers a such that there exists a positive integer n and sets A_1, A_2, $\cdots$, A_n satisfying the following conditions:

(1) every set A_i has infinitely many elements;

(2) every pair of distinct sets A_i and A_j do not share any common element;

(3) the union of sets A_1, A_2, $\cdots$, A_n is the set of all integers;

(4) for every set A_i, the positive difference of any pair of elements in A_i is at least a^i. (posed by Yuan Hanhui)

Solution The answer of the problem is the set of all positive real numbers less than 2. We consider two cases.

Case I We assume that $0 < a < 2$. Then there is a positive n such that $2^{n-1} > a^n$. We define $A_n = \{m \mid m \text{ is a multiple of } 2^{n-1}\}$ and

$$A_i = \{2^{i-1}m \mid m \text{ is an odd integer}\},$$

for $1 \leq i \leq n-1$. Then A_1, A_2, $\cdots$, A_n is a partition of the set of positive integers satisfying the conditions of the problem.

Case II We assume that $a \geq 2$. We claim that no such partition exists. To prove by contradiction, we assume on the contrary that A_1, A_2, $\cdots$, A_n is a partition satisfying the conditions of the problem. Let $N = \{1, 2, \cdots, 2^n\}$. For every i with $1 \leqslant i \leqslant n$, let $B_i = A_i \cap N$. We assume that $B_i = \{b_1, b_2, \cdots, b_m\}$ with $b_1 < b_2 < \cdots < b_m$. We have

$$2^n > b_m - b_1 = (b_m - b_{m-1}) + (b_{m-1} - b_{m-2})$$
$$+ \cdots + (b_2 - b_1) \geqslant (m-1)2^i,$$

implying that $m - 1 < 2^{n-i}$, or $m \leqslant 2^{n-i}$. Since $A_1, A_2, \cdots, A_n$ is a partition, $B_i \cap B_j = \varnothing$ for $1 \leqslant i < j \leqslant n$ and $N = B_1 \cup B_2 \cup \cdots \cup B_n$. It follows that

$$2^n = |N| = |B_1| + |B_2| + \cdots + |B_n| \leqslant \sum_{i=1}^{n} 2^{n-i} = 2^n - 1,$$

which is impossible. Hence our assumption was wrong and such a partition does not exist for every positive integer n.

Second Day
8:00 – 12:00 August 13, 2005

⬤ Let x and y be positive real numbers with $x^3 + y^3 = x - y$. Prove that $x^2 + 4y^2 < 1$. (posed by Xiong Bin)

Solution   In view of $x^3 + y^3 > 0$, it suffices to show that $(x^2 + 4y^2)(x - y) < x^3 + y^3$. Expanding the left-hand side of the last inequality and canceling the like terms we obtain $4xy^2 < x^2y + 5y^3$. By the AM-GM inequality, we have $x^2y + 5y^3 \geqslant 2\sqrt{5}xy^2 > 4xy^2$.

⬤ An integer n is called good if $n \geqslant 3$ and there are n lattice points $P_1, P_2, \cdots, P_n$ in the coordinate plane satisfying the following conditions: If line segment P_iP_j has a rational length, then there is P_k such that both line segments P_iP_k and P_jP_k have irrational lengths; and if line segment P_iP_j has an irrational length, then there is P_k such that both line segments P_iP_k and P_jP_k have rational lengths.

(1) Determine the minimum good number.

(2) Determine if 2005 is a good number.

(A point in the coordinate plane is a lattice point if both of its coordinate are integers.) (posed by Feng Zuming)

Solution   We claim that the minimum good number is 5, and that 2 005 is good.

We say P_iP_j is rational (irrational) if segment P_iP_j has a rational (an irrational) length. We say an ordered triple (P_i, P_j, P_k) of lattice points to be good with $1 \leqslant i < j \leqslant n$ and $1 \leqslant k \leqslant n$ if P_iP_j is rational (irrational) and both P_iP_k and P_jP_k are irrational (rational). In the figure shown below, if P_iP_j is rational (irrational) then P_i and P_j are connected by darkened solid (regular solid) line segment.

It is not difficult to see that $n = 3$ is not a good number. Note that $n = 4$ is also not a good number. Assume on the contrary that there are lattice points P_1, P_2, P_3, P_4 satisfying the conditions of the problem. Without loss of generality, we assume that P_1P_2 is rational and (P_1, P_2, P_3) is good. Then (P_2, P_3, P_4) must be good. Neither (P_2, P_4, P_1) nor (P_2, P_4, P_3) is good, violating the condition of the problem. Hence our assumption was wrong and $n = 4$ is not a good number.

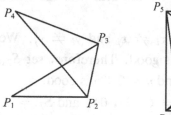 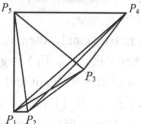

For $n = 5$, we define the set

$$S_5 = \{P_1 = (0, 0), P_2 = (1, 0), P_3 = (5, 3),$$

$$P_4 = (8, 7), P_5 = (0, 7)\}.$$

It is not difficult to see that all triples (P_i, P_j, P_k) are good. Hence $n = 5$ is minimum good number.

Consider the sets

$$A = \{(1, 0), (2, 0), \cdots, (669, 0)\},$$

$$B = \{(1, 1), (2, 1), \cdots, (668, 1)\},$$

$$C = \{(1, 2), (2, 2), \cdots, (668, 2)\}.$$

We claim that the 2 005 points in the set $S_{2\,005} = A \cup B \cup C$ satisfies the conditions of the problem. Consider a pair of distinct points $P_i = (x_i, y_i)$ and $P_j = (x_j, y_j)$ in the set $S_{2\,005}$. If either $x_i = x_j$ or $y_i = y_j$, then clearly $P_i P_j$ is rational. Otherwise, either $|x_i - x_j| = 1$ or 2, or $|y_i - y_j| = 1$ or 2, or both. Hence $|P_i P_j|^2$ can be written in the form of either $n^2 + 1$ or $n^2 + 4$ for some positive integer n. For a pair of positive integers n and m with $n > m$, we have $n^2 - m^2 \geqslant n^2 - (n-1)^2 = 2n - 1$. It follows that for a positive integer n, $n^2 + 1$ and $n^2 + 4$ are not squares of an integer. We conclude that (P_i, P_j) is rational if and only if line $P_i P_j$ is parallel to one of the axes.

If $P_i P_j$ is rational with $y_i = y_j$, then we set $P_k = (x_k, y_k)$, with $x_k \neq x_i$, $x_k \neq x_j$ and $y_k \neq y_i$. (We can do so because there are 668 distinct x values and three distinct y values.) Then (P_i, P_j, P_k) is good. In exactly the same way, we can deal with the case when $P_i P_j$ is rational with $x_i = x_j$.

If $P_i P_j$ is irrational, then $x_i \neq x_j$ and $y_i \neq y_j$. We set $P_k = (x_i, y_j)$. Then (P_i, P_j, P_k) is good. Therefore, set $S_{2\,005}$ satisfies the conditions of the problem and $n = 2\,005$ is good.

Remark Set $S_6 = S_5 \cup \{P_6 = (-24, 0)\}$ and $S_7 = S_6 \cup \{P_7 = (-24, 7)\}$. It is not difficult to see that both 6 and 7 are good. For every positive integer $n \geqslant 8$, we can easily generalize the construction of part (2) to show that n is good. Hence all integers greater than 4 are good.

Let m and n be positive integers with $m > n \geqslant 2$. Set $S = \{1, 2, \cdots, m\}$, and $T = \{a_1, a_2, \cdots, a_n\}$ is a subset of S such that every number in S is not divisible by any two distinct numbers in T. Prove that

$$\frac{1}{a_1} + \frac{1}{a_2} + \cdots + \frac{1}{a_n} < \frac{m+n}{m}.$$

(posed by Zhang Tongjun)

Solution For every i with $1 \leqslant i \leqslant n$, we define set

$$S_i = \{b \mid b \text{ is an element in } S \text{ and is divisible by } a_i\}.$$

There are $\left[\dfrac{m}{a_i}\right]$ elements in S_i. Since every element in S is not divisible by any two distinct elements in T, it follows that $S_i \cap S_j = \varnothing$ for $1 \leqslant i < j \leqslant n$. Thus

$$\sum_{i=1}^{n} \left[\frac{m}{a_i}\right] = \sum_{i=1}^{n} |S_i| \leqslant |S| = m.$$

Note that $\dfrac{m}{a_i} < \left[\dfrac{m}{a_i}\right] + 1$. It follows that

$$\sum_{i=1}^{n} \frac{m}{a_i} < \sum_{i=1}^{n} \left(\left[\frac{m}{a_i}\right] + 1\right) = \sum_{i=1}^{n} \left[\frac{m}{a_i}\right] + \sum_{i=1}^{n} 1 \leqslant m + n.$$

So the desired result is obtained.

⑤ Given an $a \times b$ rectangle with $a > b > 0$, determine the minimum length of a square that covers the rectangle. (A square covers the rectangle if each point in the rectangle lies inside the square.) (posed by Chen Yonggao)

Solution Let R denote the rectangle, and let S denote the square with minimum length that covers R. Let s denote the length of a side of S. We claim that R is inscribed in S, that is, the vertices of R lie on the sides of S. We also claim that R can only be inscribed in two ways, as shown below. Let S_1 (S_2) denote the square shown on the left-hand side (right-hand side). For S_2, the sides of R are parallel to the diagonals of S_2. It easy to see that $s = a$ if $S = S_1$.

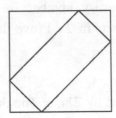

It is not difficult to see that $s = \dfrac{\sqrt{2}(a+b)}{2}$ if $S = S_2$. Taking the minimum value of $\left\{ a, \dfrac{\sqrt{2}(a+b)}{2} \right\}$, we conclude that

$$s = \begin{cases} a, & \text{if } a < (\sqrt{2}+1)b, \\ \dfrac{\sqrt{2}(a+b)}{2}, & \text{if } a \geqslant (\sqrt{2}+1)b. \end{cases}$$

Now we prove our claim that these are only two ways. Let $R = ABCD$ and $S = XYZW$. Without loss of generality, we place XY horizontally. By the minimality of S, we can assume that at least one vertex, say A, of R lies on one side of S, say WX (see the left-hand side figure shown below). If neither B nor D lies on the sides of S, we can then slide R down (vertically), so that one of them, say B lies on side XY (see the middle figure shown below). If neither C or D lies on the sides of S, then we can apply an enlargement, centered at X with scale less than 1, to S such that the image of S still covers R. This violates the minimality of S. Hence at least one of C and D lies on the sides of S, that is, three consecutive vertices of R lie on the sides of S (see the right-hand side figure shown below). Without loss of generality, we assume that they are A, B and C. If any of these three vertices coincide with any of the vertices of S, then we clearly have $S = S_1$. Hence we may assume that A, B and C are on sides WX, XY and YZ, respectively. By symmetry, we may also assume that $AB = a > b = BC$.

If D does not lie on line segment ZW, then we can slide R up a bit so both B and D lie in the interior of S (see the left-hand slide

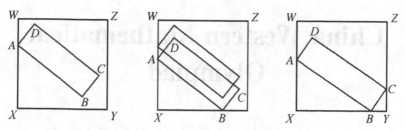

figure shown below). Let O be the center of R. We can then rotate R around O with a small angle so that all four vertices lie inside S (see the middle figure shown below). It is easy to see that we can use a smaller square to cover R (by applying an enlargement centered at O with a scale less that 1), violating the minimality of S. Thus our assumption was wrong, and D must lie on side ZW (see the right-hand side figure shown below), which is the case when $S = S_2$.

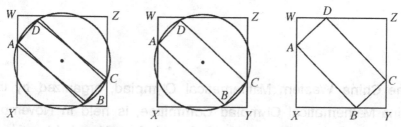

We finish our proof that if $S = S_2$, then the sides of R are parallel to the diagonals of S_2. By symmetry, it suffices to show that $AX = XB$. It is not difficult to see that $\angle XAB = \angle YBC = \angle ZCD = \angle WDA$. So triangles ABX, BCY, CDZ and DAW are similar. Set $AX = ax$ and $XB = ay$. Then $BY = bx$ and $CY = by$. Also, $DW = bx$ and $WA = by$. Hence $by + ax = WA + AX = WX = XY = XB + BY = ay + bx$, implying that $(a-b)x = (a-b)y$, or $x = y$, as desired.

China Western Mathematical Olympiad

The China Western Mathematical Olympiad, organized by the China Mathematical Olympiad Committee, is held in November every year. Students of grade 10 and 11, from Kasakshtan, Hong Kong, Macau and West China, are invited to take part in the competition. The competition lasts for 2 days, and there are 4 problems to be completed within 4 hours each day.

2002 (Lanzhou, Gansu)

The 2nd (2002) China Western Mathematical Olympiad was held on December 3 – 8, 2002 in Lanzhou, Gansu, China, and was hosted by Gansu Mathematical Society and Lanzhou University.

The Competition Committee consisted of the following: Pan Chenbiao, Fan Xianling, Li Shenghong, Leng Gangsong, Xiong Bin, Feng Zhigang, Wang Haiming and Zhao Dun.

First Day
8:00 – 12:00 December 4, 2002

1 Find all positive integers n such that

$$n^4 - 4n^3 + 22n^2 - 36n + 18$$

is a perfect square. (posed by Pan Chengbiao)

Solution We write

$$A = n^4 - 4n^3 + 22n^2 - 36n + 18$$
$$= (n^2 - 2n)^2 + 18(n^2 - 2n) + 18.$$

Let $n^2 - 2n = x$, $A = y^2$, where y is a nonnegative integer. Then

$$(x+9)^2 - 63 = y^2.$$

That is, $(x+9-y)(x+9+y) = 63$.

It can only be $(x+9-y, x+9+y) = (1, 63), (3, 21)$ or $(7, 9)$. We obtain, respectively, $(x, y) = (23, 31), (3, 9)$ or $(-1, 1)$. But only when $x = 3$ and -1, $n^2 - 2n = x$ has solutions in positive integers and we get $n = 1$ or 3.

Therefore, $n = 1$ or 3 satisfies the condition.

Remark We can also deal with this problem by using inequality.

2 Suppose O is the circumcenter of an acute triangle $\triangle ABC$, P is a point inside $\triangle AOB$, and D, E, F are the projections of P on three sides BC, CA, AB of $\triangle ABC$ respectively. Prove that a parallelogram with FE and FD as adjacent sides lies inside $\triangle ABC$. (posed by Leng Gangsong)

Proof As shown in the figure, we construct a parallelogram $DEFG$ with FE and FD as adjacent sides. To prove the proposition to be true, we need only to prove that $\angle FEG < \angle FEC$, and $\angle FDG < \angle FDC$. It is equivalent to proving: $\angle BFD < \angle BAC$, and $\angle AFE < \angle ABC$.

In fact, we construct OH with $OH \perp BC$, and H is the foot of the perpendicular. From $PD \perp BC$ and $PF \perp AB$, we know that four points B, F, P and D are concyclic. Thus $\angle BFD = \angle BPD$. But $\angle PBD > \angle OBH$, hence $90° - \angle PBD < 90° - \angle OBH$, and that is, $\angle BPD < \angle BOH$. Moreover, O is the circumcenter of $\triangle ABC$, so $\angle BOH = \frac{1}{2}\angle BOC = \angle BAC$. Therefore, $\angle BFD = \angle BPD < \angle BOH = \angle BAC$, that is, $\angle BFD < \angle BAC$.

Similarly, we can prove that $\angle AFE < \angle ABC$. Therefore, the proposition holds.

Remark If O is not the circumcenter of $\triangle ABC$, but the incenter or orthocenter, does the conclusion given in the problem still hold? The reader may wish to think over it.

⬤ Consider a square on the complex plane. The complex numbers corresponding to its four vertices are the four roots of some equation of the fourth degree with one unknown and integer coefficients $x^4 + px^3 + qx^2 + rx + s = 0$. Find the minimum value of the area of such square. (posed by Xiong Bin)

Solution Suppose the complex number corresponding to the center of the square is a. Then after translating the origin of the complex plane to a, the vertices of the square distribute evenly on the circumference. That is, they are the solutions of equation $(x-a)^4 = b$, where b is a complex number. Hence,

$$x^4 + px^3 + qx^2 + rx + s = (x-a)^2 - b$$
$$= x^4 - 4ax^3 + 6a^2x^2 - 4a^3x + a^4 - b.$$

Comparing the coefficients of terms for x with the same degree, we know that $-a = \dfrac{p}{4}$, and it is a rational number. Combining further that $-4a^3 = r$ is an integer, we can see that a is an integer. So by using the fact that $s = a^4 - b$ is an integer, we can show that b is also an integer.

The above discussion makes clear of a fact that the four numbers corresponding to the four vertices of this square are roots of integer coefficients equation $(x - a)^4 = b$. Hence, the radius of its circumcircle $(= \sqrt[4]{|b|})$ is not less that 1. Therefore, the area of this square is not less than $(\sqrt{2})^2 = 2$. But the four roots of the equation $x^4 = 1$ are corresponding to the four vertices of a square on the complex plane. Hence, the minimum value of the area of the square is 2.

Remark　By using the method above, we can prove: If the corresponding complex numbers of vertices of a regular n-gon are n complex roots of some equation with integer coefficients $x^n + a_{n-1}x^{n-1} + \cdots + a_0 = 0$, then the minimum value of the area of a regular n-gon is $\dfrac{n}{2}\sin\dfrac{2\pi}{n}$.

　　Assume that n is a positive integer, and $A_1, A_2, \cdots, A_{n+1}$ are $n+1$ nonempty subsets of the set $\{1, 2, \cdots, n\}$. Prove that there are two disjoint and nonempty subsets $\{i_1, i_2, \cdots, i_k\}$ and $\{j_1, j_2, \cdots, j_m\}$ such that

$$A_{i_1} \cup A_{i_2} \cup \cdots \cup A_{i_k} = A_{j_1} \cup A_{j_2} \cup \cdots \cup A_{j_m}.$$

(posed by Zhao Dun)

Proof I　We prove by induction for n.

When $n = 1$, $A_1 = A_2 = \{1\}$, the proposition holds.

Suppose it holds for n. We consider the case of $n+1$.

Suppose $A_1, A_2, \cdots, A_{n+2}$ are nonempty subsets of $\{1, 2, \cdots, n+1\}$. Let $B_i = A_i \backslash \{n+1\}$, $i = 1, 2, \cdots, n+2$. We will prove the following cases.

Case I　There exist $1 \leqslant i < j \leqslant n+2$ such that $B_i = B_j = \varnothing$, then $A_i = A_j = \{n+1\}$. The proposition is proven.

Case II   There exists only one i such that $B_i = \varnothing$. There is no loss of generality in supposing $B_{n+2} = \varnothing$, and that is $A_{n+2} = \{n+1\}$. Now by the inductive assumption, for $\{1, 2, \cdots, n+1\}$, there exist two disjoint subsets $\{i_1, \cdots, i_k\}$ and $\{j_1, \cdots, j_m\}$ such that

$$B_{i_1} \cup \cdots \cup B_{i_k} = B_{j_1} \cup \cdots \cup B_{j_m}. \qquad \text{①}$$

We write $C = A_{i_1} \cup \cdots \cup A_{i_k}$, $D = A_{j_1} \cup \cdots \cup A_{j_m}$. Then C and D differ at the most by the element $n+1$. (This can be shown by ① and the definition of B_i.) In this case, we can make the proposition to hold true by putting A_{n+2} into C or D.

Case III   No B_i is empty. Now B_1, B_2, $\cdots$, B_{n+1} are nonempty subsets of $\{1, 2, \cdots, n\}$. By the inductive assumption, we show that, for $\{1, 2, \cdots, n+1\}$, there exist disjoint subsets $\{i_1, \cdots, i_k\}$ and $\{j_1, \cdots, j_m\}$ such that

$$B_{i_1} \cup \cdots \cup B_{i_k} = B_{j_1} \cup \cdots \cup B_{j_m}. \qquad \text{②}$$

In addition, B_2, B_3, $\cdots$, B_{n+2} are also nonempty subsets of $\{1, 2, \cdots, n\}$. By the inductive assumption, we can show that, for $\{2, 3, \cdots, n+2\}$, there exist disjoint subsets $\{r_1, \cdots, r_u\}$ and $\{t_1, \cdots, t_v\}$ such that

$$B_{r_1} \cup \cdots \cup B_{r_u} = B_{t_1} \cup \cdots \cup B_{t_v}. \qquad \text{③}$$

Again, we write $C = A_{i_1} \cup \cdots \cup A_{i_k}$, $D = A_{j_1} \cup \cdots \cup A_{j_m}$, and write $E = A_{r_1} \cup \cdots \cup A_{r_u}$, $F = A_{t_1} \cup \cdots \cup A_{t_v}$. By using ② , ③ and the definition of B_i, we see that C and D differ at the most by the element $n+1$, and so do E and F. If $C = D$ or $E = F$, then the proposition holds. Hence we need only to consider the case when $C \neq D$ and $E \neq F$. There is no loss of generality in supposing $C = D \cup \{n+1\}$, but $E = F \backslash \{n+1\}$. Now $C \cup E = D \cup F$. After amalgamating the sets occurred repeatedly in C and E, as well as in D and F, we get two subsets $\{p_1, \cdots, p_x\}$ and $\{q_1, \cdots, q_y\}$ of $\{1, 2, \cdots, n+2\}$ such that

$$A_{p_1} \cup \cdots \cup A_{p_x} = A_{q_1} \cup \cdots \cup A_{q_y}, \qquad ④$$

where $G = A_{p_1} \cup \cdots \cup A_{p_x} = C \cup E$, $H = A_{q_1} \cup \cdots \cup A_{q_y} = D \cup F$.

Now, if $\{p_1, \cdots, p_x\} \cap \{q_1, \cdots, q_y\} = \varnothing$, then the proposition holds. If there is $i \in \{p_1, \cdots, p_x\} \cap \{q_1, \cdots, q_y\}$, we write $\widetilde{C} = \{A_{i_1}, \cdots, A_{i_k}\}$, $\widetilde{D} = \{A_{j_1}, \cdots, A_{j_m}\}$, $\widetilde{E} = \{A_{r_1}, \cdots, A_{r_u}\}$, $\widetilde{F} = \{A_{t_1}, \cdots, A_{t_v}\}$. And there is no loss of generality in assuming that A_i does not belong to $\widetilde{C}$ and $\widetilde{E}$ at the same time, and it does not belong to $\widetilde{D}$ and $\widetilde{F}$ at the same time too. Hence there are only two possibilities.

(a) $A_i \in \widetilde{C}$ and $A_i \in \widetilde{F}$. If there are two sets in $\widetilde{C}$ containing $n + 1$, then we take away set A_i from the left side in ④. Now since all elements except $n+1$ in A_i belong to E (in view of ③), and there are two sets on the left side in ④ containing $n+1$. Thus after taking away A_i, the number of elements in G does not reduce and ④ is still an equality. In the same way, if there are two sets in $\widetilde{F}$ containing $n+1$, then we take away A_i from the right side in ④, and ④ still holds.

Of course, if there is only one set in $\widetilde{C}$ and $\widetilde{F}$ containing $n + 1$, then after taking away A_i from both sides in ④, it remains to be an equality. (Now, by ② and ③, we can see that the two sides of ④ will not become empty sets.)

(b) $A_i \in \widetilde{D}$ and $A_i \in \widetilde{E}$, then $n+1 \notin A_i$. Now after taking away A_i from both sides in ④, the resulting expression is still an equality.

In view of the above operation, we have a method to make the two sets of subscripts $\{p_1, \cdots, p_x\}$ and $\{q_1, \cdots, q_y\}$ in ④ disjoint. Therefore the proposition holds for $n + 1$.

As a consequence of what is described above, we show that the proposition holds.

Proof II Here we need to use a fact from linear algebra that $n + 1$ vectors in the n-dimensional linear space are linearly dependent.

If element i is in set A_j, we write it as 1, otherwise write it as 0. Then A_j corresponds to an n-dimensional vector, which is nonzero and contains 0 and 1. We write $\alpha_j = (a_{j_1}, a_{j_2}, \cdots, a_{j_n})$, where

$$a_{j_i} = \begin{cases} 1, & i \in A_j, \\ 0, & i \notin A_j. \end{cases}$$

Since a_1, a_2, $\cdots$, a_{n+1} are $n+1$ vectors in the n-dimensional space, so there exists a group of real numbers, not every one of them to be zero, x_1, x_2, $\cdots$, x_{n+1} such that

$$x_1 a_1 + x_2 a_2 + \cdots + x_{n+1} a_{n+1} = 0. \qquad \text{⑤}$$

Hence, for $\{1, 2, \cdots, n+1\}$, there exist two disjoint and nonempty subsets $\{i_1, \cdots, i_k\}$ and $\{j_1, \cdots, j_m\}$ such that

$$x_{i_1} a_{i_1} + \cdots + x_{i_k} a_{i_k} = y_{j_1} a_{j_1} + \cdots + y_{j_m} a_{j_m}, \qquad \text{⑥}$$

where $x_{i_1}, \cdots, x_{i_k} > 0$, $y_{j_1} = (-x_{j_1})$, $\cdots$, $y_{j_m} = (-x_{j_m}) > 0$ (Here, it is essential to put the terms with coefficients greater than zero in ⑤ to one side, and those with coefficients less than zero to another side).

We conclude that

$$A_{i_1} \cup A_{i_2} \cup \cdots \cup A_{i_k} = A_{j_1} \cup \cdots \cup A_{j_m}. \qquad \text{⑦}$$

In fact, if element a $(1 \leqslant a \leqslant n)$ belongs to the left side in ⑦, then the a-th component of the sum of the vectors from the left side in ⑥ must be greater than zero. Thus it makes the a-th component of the sum of the vectors from the right side in ⑥ to be greater than zero. Hence, there is a_{j_t}, and its a-th component is 1, that is, $a \in A_{j_t}$. Conversely, it is also true, that is, ⑦ holds.

Therefore, the original proposition holds.

Remark Proof Ⅱ is easy and fundamental. But it needs to use some knowledge of linear algebra. Although such knowledge is elementary, but it is not faught in the middle schools. In Proof Ⅰ, after taking away $n+1$ and obtaining conclusion by using inductive assumption, we may not put $n+1$ back simply. Professor Pan Chengbiao obtains this proof after putting in a lot of hard work.

Second Day
8:00 – 12:00 December 5, 2002

5 In a given trapezium $ABCD$, $AD /\!/ BC$. Suppose E is a variable point on AB, O_1 and O_2 are circumcenters of $\triangle AED$ and $\triangle BEC$ respectively. Prove that the length of O_1O_2 is a fixed value. (posed by Leng Gangsong)

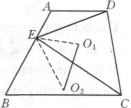

Proof As shown in the figure, we join EO_1 and EO_2, then $\angle AEO_1 = 90° - \angle ADE$, $\angle BEO_2 = 90° - \angle BCE$. Hence

$$\angle O_1EO_2 = \angle ADE + \angle ECB.$$

Since $AD /\!/ BC$, through E constructing a line parallel to AD, we can prove $\angle DEC = \angle ADE + \angle BCE$, so $\angle O_1EO_2 = \angle DEC$.

Further, by the sine rule, we can show

$$\frac{DE}{EC} = \frac{2O_1E\sin A}{2O_2E\sin B} = \frac{O_1E}{O_2E}.$$

Thus $\triangle DEC \backsim \triangle O_1EO_2$. Therefore

$$\frac{O_1O_2}{DC} = \frac{O_1E}{DE} = \frac{O_1E}{2O_1E\sin A} = \frac{1}{2\sin A}.$$

So $O_1O_2 = \dfrac{DC}{2\sin A}$, which is a fixed value. The proposition is proven.

Remark It is natural to think that the distance between the projections of O_1 and O_2 on AB is $\dfrac{1}{2}AB$, which is a fixed value. Thus we need only to prove the included angle between O_1O_2 and AB is a fixed value. This is an idea for another method to prove the problem. The reader may wish to try it.

6 Assume that n is a given positive integer. Find all of the integer groups $(a_1, a_2, \cdots, a_n)$ satisfying the conditions:

(1) $a_1 + a_2 + \cdots + a_n \geqslant n^2$;

(2) $a_1^2 + a_2^2 + \cdots + a_n^2 \leqslant n^3 + 1$.

(posed by Pan Chengbiao)

Solution   Suppose $(a_1, a_2, \cdots, a_n)$ is an integer group which satisfies the conditions. Then by Cauchy's inequality we have

$$a_1^2 + \cdots + a_n^2 \geqslant \frac{1}{n}(a_1 + \cdots + a_n)^2 \geqslant n^3. \qquad \text{①}$$

Combining $a_1^2 + \cdots + a_n^2 \leqslant n^3 + 1$, we see that it can only be $a_1^2 + \cdots + a_n^2 = n^3$ or $a_1^2 + \cdots + a_n^2 = n^3 + 1$.

If it is the former, then by the condition for Cauchy's inequality to take the equality sign, we can obtain $a_1 = \cdots = a_n$. This requires $a_i^2 = n^2$, $1 \leqslant i \leqslant n$. Combining $a_1 + \cdots + a_n \geqslant n^2$, we have $a_1 = \cdots = a_n = n$.

If it is the latter, then let $b_i = a_i - n$, then we have

$$b_1^2 + b_2^2 + \cdots + b_n^2 = \sum_{i=1}^{n} a_i^2 - 2n \sum_{i=1}^{n} a_i + n^3$$

$$= 2n^3 + 1 - 2n \sum_{i=1}^{n} a_i \leqslant 1.$$

Thus b_i^2 can only be 0 or 1, and there is at the most one among b_1^2, b_2^2, $\cdots$, b_n^2 to be 1. If b_1^2, $\cdots$, b_n^2 all are zero, then $a_i = n$, $\sum\limits_{i=1}^{n} a_i^2 = n^3 \neq n^3 + 1$. It leads to a contradiction. If there is just one among b_1^2, $\cdots$, b_n^2 to be 1, then $\sum\limits_{i=1}^{n} a_i^2 = n^3 \pm 2n + 1 \neq n^3 + 1$. Again, it leads to a contradiction.

Consequently, we obtain that it can only be $(a_1, \cdots, a_n) = (n, \cdots, n)$.

Assume that α, β are two roots of the equation $x^2 - x - 1 = 0$. Let

$$a_n = \frac{\alpha^n - \beta^n}{\alpha - \beta}, \quad n = 1, 2, \cdots.$$

(1) Prove that for any positive integer n, we have $a_{n+2} = a_{n+1} + a_n$.

(2) Find all positive integers a and b, $a < b$, satisfying that b divides $a_n - 2n \cdot a^n$ for any positive integer n. (posed by Li Shenghong)

Solution (1) Noting $\alpha + \beta = 1$ and $\alpha\beta = -1$, we have

$$\alpha^{n+2} - \beta^{n+2} = (\alpha + \beta)(\alpha^{n+1} - \beta^{n+1}) - \alpha\beta(\alpha^n - \beta^n)$$
$$= (\alpha^{n+1} - \beta^{n+1}) + (\alpha^n - \beta^n).$$

Dividing two sides by $\alpha - \beta$, we have $a_{n+2} = a_{n+1} + a_n$.

(2) By assumption, we have $b \mid a_1 - 2a$, that is, $b \mid 1 - 2a$. But $b > a$, so $b = 2a - 1$. Moreover, for an arbitrary positive integer n, we have

$$b \mid a_n - 2na^n, \quad b \mid a_{n+1} - 2(n+1)a^{n+1}, \quad b \mid a_{n+2} - 2(n+2)a^{n+2}.$$

Combining $a_{n+2} = a_{n+1} + a_n$ and $b = 2a - 1$ an odd number, we have

$$b \mid (n+2)a^{n+2} - (n+1)a^{n+1} - na^n.$$

But $(b, a) = 1$, so

$$b \mid (n+2)a^2 - (n+1)a - n. \tag{①}$$

Taking n as $n+1$ in ①, we have

$$b \mid (n+3)a^2 - (n+2)a - (n+1). \tag{②}$$

Subtracting the right side of ① from the same side of ②, we have

$$b \mid a^2 - a - 1,$$

that is,

$$2a - 1 \mid a^2 - a - 1,$$

so

$$2a - 1 \mid 2a^2 - 2a - 2.$$

But

$$2a^2 \equiv a \pmod{2a - 1}.$$

Therefore

$$2a - 1 \mid -a - 2, \quad 2a - 1 \mid -2a - 4.$$

So

$$2a - 1 \mid -5, \quad 2a - 1 = 1 \text{ or } 5.$$

But $2a - 1 = 1$ implies $b = a$, a contradiction. So $2a - 1 = 5$, $a = 3$ and $b = 5$.

We will prove in the following that when $a = 3$ and $b = 5$, for any positive integer n, we have $b \mid a_n - 2na^n$, that is $5 \mid a_n - 2n \times 3^n$.

When $n = 1$, 2, since $a_1 = 1$, $a_2 = \alpha + \beta = 1$, we have $a_1 - 2 \times 3 = -5$, $a_2 - 2 \times 2 \times 3^2 = -35$. So, when $n = 1$, 2, we have $5 \mid a_n - 2n \times 3^n$.

Assume that, when $n = k, k+1$, this conclusion is true, that is,

$$5 \mid a_k - 2k \times 3^k, \ 5 \mid a_{k+1} - 2(k+1) \times 3^{k+1}.$$

Hence $5 \mid (a_{k+1} + a_k) - 2k \times 3^k - 2(k+1)3^{k+1}$.

That is, $5 \mid a_{k+2} - 2 \times 3^k(k + 3(k+1))$,

so $5 \mid a_{k+2} - 2 \times 3^k \times (4k + 3)$.

To prove $5 \mid a_{k+2} - 2(k+2) \times 3^{k+2}$, we need only to prove

$$2(k+2) \times 3^{k+2} \equiv 2 \times (4k + 3) \times 3^k \pmod 5. \qquad \text{③}$$

It is equivalent to $9(k+2) \equiv 4k+3$, that is, $5k + 15 \equiv 0 \pmod 5$. Obviously the latter holds, Hence ③ holds. Thus the conclusion holds for $n = k + 2$.

Consequertly, we show that $(a, b) = (3, 5)$ satisfying the conditions.

Remark A sequence of numbers $\{a_n\}$ occurring in the problem is a Fibonacci sequence. This problem came from a discussion with regard to the properties of Fibonacci sequence. Problems concerning Fibonacci sequence often appear in mathematical contests.

⬤ Assume that $S = (a_1, a_2, \cdots, a_n)$ consists of 0 and 1 and is the longest sequence of number, which satisfies the following condition: Every two sections of successive 5 terms in the sequence of numbers S are different, i.e., for arbitrary $1 \leqslant i < j \leqslant n - 4$, $(a_i, a_{i+1}, a_{i+2}, a_{i+3}, a_{i+4})$ and $(a_j, a_{j+1}, a_{j+2}, a_{j+3}, a_{j+4})$ are different. Prove that the first four terms and the last four terms in the sequence are the same. (posed by Feng Zhigang)

Proof Noting that S is the longest sequence of numbers satisfying the condition. Hence, if we add a term, 0 or 1, after the last term of S, there will occur two identical sections of successive 5 terms in S, and that is, there exist $i \neq j$ such that

$$(a_i, a_{i+1}, \cdots, a_{i+4}) = (a_{n-3}, a_{n-2}, \cdots, a_n, 0),$$
$$(a_j, a_{j+1}, \cdots, a_{j+4}) = (a_{n-3}, a_{n-2}, \cdots, a_n, 1).$$

If $(a_1, a_2, a_3, a_4) \neq (a_{n-3}, a_{n-2}, a_{n-1}, a_n)$, then

$$1 < i, j < n-3, i \neq j,$$

and

$$(a_i, a_{i+1}, a_{i+2}, a_{i+3}) = (a_j, a_{j+1}, a_{j+2}, a_{j+3})$$
$$= (a_{n-3}, a_{n-2}, a_{n-1}, a_n).$$

Now, consider a_{i-1}, a_{j-1} and a_{n-4}, among which there must be two identical terms. This causes two sections with successive 5 terms respectively in S to be identical. It leads to a contradiction.

Therefore, the proposition holds.

Remark Note that there are at the most 32 different sections of successive 5 terms (consisting of 0 and 1). According to this, we know that the length for S is less than or equal to 36. The reader may wish to try to construct one such S with length 36 and satisfying the requirements.

2003 (Urumqi, Xinjiang)

The 3rd (2003) China Western Mathematical Olympiad was held on September 25 – 30, 2003 in Urumqi, Xinjiang, China, and was hosted by the Research Center of Education of Urumqi.

The Competition Committee consisted of the following: Chen

Yong-gao, Li Sheng-hong, Qiu Zong-hu, Leng Gang-song, Xiong Bin, Feng Zhi-gang, Zhang Zheng-jie.

First Day

9:30 – 13:30 September 27, 2003

Put numbers 1, 2, 3, 4, 5, 6, 7 and 8 at the vertices of a cube, such that the sum of any three numbers on any face is not less than 10. Find the minimum sum of the four numbers on a face. (posed by Qiu Zonghu)

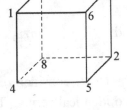

Solution Suppose that the four numbers on a face of the cube is a_1, a_2, a_3, a_4 such that their sum reaches the minimum and $a_1 < a_2 < a_3 < a_4$. Since the maximum sum of any three numbers less than 5 is 9, we have $a_4 \geqslant 6$, and then $a_1 + a_2 + a_3 + a_4 \geqslant 16$.

As seen in the figure, we have $2 + 3 + 5 + 6 = 16$, and that means the minimum sum of the four numbers on a face is 16.

Let a_1, a_2, $\cdots$, a_{2n} be real numbers with $\sum\limits_{i=1}^{2n-1}(a_{i+1} - a_i)^2 = 1$. Find the maximum value of $(a_{n+1} + a_{n+2} + \cdots + a_{2n}) - (a_1 + a_2 + \cdots + a_n)$. (posed by Leng Gangsong)

Solution First, for $n = 1$, we have $(a_2 - a_1)^2 = 1$, $a_2 - a_1 = \pm 1$. Then the maximum value of $a_2 - a_1$ is 1.

Secondly, for $n \geqslant 2$, let $x_1 = a_1$, $x_{i+1} = a_{i+1} - a_i$, $i = 1, 2, \cdots$, $2n - 1$. Then $\sum\limits_{i=2}^{2n} x_i^2 = 1$, and $a_k = x_1 + \cdots + x_k$, $k = 1, 2, \cdots, 2n$. Using Cauchy's Inequality, we have

$$(a_{n+1} + a_{n+2} + \cdots + a_{2n}) - (a_1 + a_2 + \cdots + a_n)$$

$$= n(x_1 + \cdots + x_n) + nx_{n+1} + (n-1)x_{n+2} + \cdots + x_{2n}$$

$$- [nx_1 + (n-1)x_2 + \cdots + x_n]$$

$$= x_2 + 2x_3 + \cdots + (n-1)x_n + nx_{n+1} + (n-1)x_{n+2} + \cdots + x_{2n}$$

$$\leqslant [1^2 + 2^2 + \cdots + (n-1)^2 + n^2 + (n-1)^2 + \cdots + 1^2]^{\frac{1}{2}} (x_2^2$$

$$+ x_3^2 + \cdots + x_{2n}^2)^{\frac{1}{2}}$$

$$= \left[n^2 + 2 \times \frac{1}{6}(n-1)n(2n-1) \right]^{\frac{1}{2}} = \sqrt{\frac{n(2n^2+1)}{3}}.$$

The equality holds when

$$a_k = \frac{\sqrt{3}k(k-1)}{2\sqrt{n(2n^2+1)}}, \quad a_{n+k} = \frac{\sqrt{3}[2n^2 - (n-k)(n-k+1)]}{2\sqrt{n(2n^2+1)}},$$

$$k = 1, 2, \cdots, n.$$

So the maximum value of $(a_{n+1} + a_{n+2} + \cdots + a_{2n}) - (a_1 + a_2 + \cdots + a_n)$ is $\sqrt{\dfrac{n(2n^2+1)}{3}}$.

Remark   The first, second and third problems in this competition are all concerning with finding the maximum or minimum values. A solution of such a kind of problems usually involves two steps: First, get an upper bound or a lower bound of the problem using given conditions and well-known inequalities. Secondly, show that such an upper bound or lower bound is attainable, and then is the maximum or minimum value we are looking for.

◉ Let n be a given positive integer. Find the least positive integer u_n, such that for any positive integer d, the number of integers divisible by d in every u_n consecutive positive odd numbers is not less than the number of integers divisible by d in $1, 3, 5, \cdots, 2n-1$. (posed by Chen Yonggao)

Solution   The correct answer is $u_n = 2n - 1$. The proof is given in the following.

(1) $u_n \geqslant 2n - 1$. As $u_1 = 1$, we only need to consider $n \geqslant 2$. Since the number of integers divisible by $2n-1$ in $1, 3, 5, \cdots, 2n-1$ is 1 and that in $2(n+1) - 1, 2(n+2) - 1, \cdots, 2(n+2n-2) - 1$ is 0, then

$u_n \geqslant 2n-1$.

(2) $u_n \leqslant 2n-1$. We only need to consider the case when $2 \nmid d$ and $1 \leqslant d \leqslant 2n-1$. For any $2n-1$ consecutive positive odd numbers: $2(a+1)-1$, $2(a+2)-1$, $\cdots$, $2(a+2n-1)-1$, let s and t be positive integers such that

$$(2s-1)d \leqslant 2n-1 < (2s+1)d,$$

$$(2t-1)d \leqslant 2(a+1)-1 < (2t+1)d.$$

Then the number of integers divisible by d in $1, 3, 5, \cdots, 2n-1$ is s, and

$$(2(t+s)-1))d = (2t-1)d + (2s-1)d + d$$
$$\leqslant 2(a+1)-1+2n-1+2n-1$$
$$= 2(a+2n-1)-1.$$

That means $u_n \leqslant 2n-1$.

Remark The key to the solution lies in finding out the fact that the number of integers divisible by $2n-1$ in $2n-2$ consecutive positive odd numbers $2(n+1)-1$, $2(n+2)-1$, $\cdots$, $2(n+2n-2)-1$ is 0, then $u_n \geqslant 2n-1$. The remaining is to prove that $u_n \leqslant 2n-1$.

Suppose the sum of distances from any point P in a convex quadrilateral $ABCD$ to lines AB, BC, CD and DA is constant. Prove that $ABCD$ is a parallelogram. (posed by Xiong Bin)

Solution Let $d(P, l)$ denote the distance from point P to line l. We first prove the following lemma.

Lemma *Let $\angle SAT = \alpha$ be a given angle and P a moving point in $\angle SAT$. If the sum of distances from P to lines AS and AT is a constant number m, then the trace of P is a segment BC with points B and C on AS and AT respectively, and $AB = AC = \dfrac{m}{\sin \alpha}$. Further, if a point Q lies inside $\triangle ABC$, then the sum of*

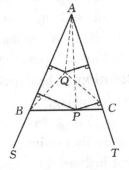

distances from Q to lines AS and AT is less than m, and it is greater than m if Q lies outside $\triangle ABC$.

Proof of Lemma   For $AB = AC = \dfrac{m}{\sin \alpha}$ and P on BC, We have

$S_{\triangle PAB} + S_{\triangle PAC} = S_{\triangle ABC}$, i.e. $\dfrac{1}{2} AB \cdot d(P, AB) + \dfrac{1}{2} AC \cdot d(P,$

$AC) = \dfrac{1}{2} \cdot AB \cdot AC \cdot \sin \alpha$. Then $d(P, AB) + d(P, AC) = m$. If

point Q lies inside $\triangle ABC$, $S_{\triangle QAB} + S_{\triangle QAC} < S_{\triangle ABC}$, then $d(Q,$

$AB) + d(Q, AC) < m$. If Q lies outside $\triangle ABC$, $S_{\triangle QAB} + S_{\triangle QAC} >$

$S_{\triangle ABC}$, then $d(Q, AB) + d(Q, AC) > m$. The proof of Lemma is complete.

We now consider the following two cases:

(1) Neither pair of the opposite sides of the quadrilateral *ABCD* is parallel. We may assume that sides *BC* and *AD* meet at point *F* and sides *BA* and *CD* meet at point *E*. Through point *P* draw segments l_1 and l_2 such

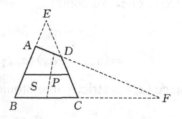

that the sum of distances from any point on l_1 to *AB* and *CD* is constant and the sum of distances from any point on l_2 to *BC* and *AD* is also constant. As seen in the second figure, for any point Q in area S, using the given condition and the lemma, we have

$$d(P, AB) + d(P, BC) + d(P, CD) + d(P, DA)$$

$$= d(Q, AB) + d(Q, BC) + d(Q, CD) + d(Q, DA)$$

$$= [d(Q, AB) + d(Q, CD)] + [d(Q, BC) + d(Q, DA)]$$

$$> [d(P, AB) + d(P, CD)] + [d(P, BC) + d(P, DA)].$$

It leads to a contradiction.

(2) The quadrilateral *ABCD* is a trapezoid. In the same way, we can also show that it will lead to a contradiction.

This completes the proof.

Second Day

9:30－13:30 September 28, 2003

⑤ Let k be a given positive integer and $\{a_n\}$ a number sequence with $a_0 = 0$ and

$$a_{n+1} = ka_n + \sqrt{(k^2-1)a_n^2 + 1}, \ n = 0, 1, 2, \cdots.$$

Prove that every term of the sequence $\{a_n\}$ is an integer and $2k \mid a_{2n}$ for all n. (posed by Zhang Zhenjie)

Solution　We have $a_{n+1}^2 - 2k a_n a_{n+1} + a_n^2 - 1 = 0$ and $a_{n+2}^2 - 2ka_{n+1}a_{n+2} + a_{n+1}^2 - 1 = 0$. Subtracting the first expression from the second one, we obtain

$$a_{n+2}^2 - a_n^2 - 2ka_{n+1}a_{n+2} + 2ka_n a_{n+1}$$

$$= (a_{n+2} - a_n)(a_{n+2} + a_n - 2ka_{n+1}) = 0.$$

Since $\{a_n\}$ is strictly monotonic increasing, then

$$a_{n+2} = 2ka_{n+1} - a_n. \qquad\qquad ①$$

From $a_0 = 0$, $a_1 = 1$ and ①, we get that every term of $\{a_n\}$ is an integer.

Furthermore, from ① we have

$$2k \mid a_{n+2} - a_n. \qquad\qquad ②$$

From $2k \mid a_0$ and ②, we get $2k \mid a_n$, $n = 1, 2, \cdots$.

Remark　Apparently, the recursive relation in $\{a_n\}$ is nonlinear. Actually it is not. When we encounter a problem involving a nonlinear sequence, the first thing to do is to linearize and to simplify the problem. Specifically, if it has an expression with radical terms, try to eliminate these terms to get a linear recursive expression of order 2.

⑥ Suppose the convex quadrilateral $ABCD$ has an inscribed circle. The circle touches AB, BC, CD, DA at A_1, B_1, C_1, D_1

respectively. Let points E, F, G, H be the midpoints of $A_1 B_1$, $B_1 C_1$, $C_1 D_1$, $D_1 A_1$ respectively. Prove that quadrilateral $EFGH$ is a rectangle if and only if $ABCD$ is a cyclic quadrilateral. (posed by Feng Zhigang)

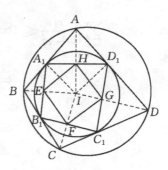

Solution   As seen in the figure, let point I be the center of the inscribed circle of $ABCD$. Since H is the midpoint of $D_1 A_1$ and lines AA_1, AD_1 are the two tangent lines through A to the circle, then point H lies on the line segment AI and $AI \perp A_1 D_1$. From $ID_1 \perp AD_1$ and using the proportional theorem for similar triangles we get that $IH \cdot IA = ID_1^2 = r^2$, where r is the radius of the inscribed circle. In the same way, we get that $IE \cdot IB = r^2$. Then $IE \cdot IB = IH \cdot IA$, and that means points A, H, E, B lie on one circle, so $\angle EHI = \angle ABE$. Similarly, we obtain that $\angle IHG = \angle ADG$, $\angle IFE = \angle CBE$, $\angle IFG = \angle CDG$. Adding these four equations, we get that $\angle EHG + \angle EFG = \angle ABC + \angle ADC$, and that means A, B, C, D lie on one circle if and only if E, F, G, H lie on one circle. Notice that quadrilateral $EFGH$ is a parallelogram as E, F, G, H are the midpoints of the sides of quadrilateral $A_1 B_1 C_1 D_1$ respectively. So E, F, G, H lie on one circle if and only if $EFGH$ is a rectangle. This completes the proof.

Let x_1, x_2, $\cdots$, x_5 be nonnegative real numbers with $\sum_{i=1}^{5} \dfrac{1}{1+x_i} = 1$.

Prove that $\sum_{i=1}^{5} \dfrac{x_1}{4+x_i^2} \leqslant 1$. (posed by Li Shenghong)

Solution   Let $y_i = \dfrac{1}{1+x_i}$, $i = 1, 2, \cdots, 5$, then $x_i = \dfrac{1-y_i}{y_i}$, $i = 1$,

$2, \cdots, 5$ and $\sum_{i=1}^{5} y_i = 1$.

We have

$$\sum_{i=1}^{5} \frac{x_i}{4+x_i^2} \leqslant 1 \Leftrightarrow \sum_{i=1}^{5} \frac{-y_i^2+y_i}{5y_i^2-2y_i+1} \leqslant 1$$

$$\Leftrightarrow \sum_{i=1}^{5} \frac{-5y_i^2+5y_i}{5y_i^2-2y_i+1} \leqslant 5$$

$$\Leftrightarrow \sum_{i=1}^{5} \left(-1+\frac{3y_i+1}{5y_i^2-2y_i+1}\right) \leqslant 5$$

$$\Leftrightarrow \sum_{i=1}^{5} \frac{3y_i+1}{5\left(y_i-\frac{1}{5}\right)^2+\frac{4}{5}} \leqslant 10.$$

Furthermore,

$$\sum_{i=1}^{5} \frac{3y_i+1}{5\left(y_i-\frac{1}{5}\right)^2+\frac{4}{5}} \leqslant \sum_{i=1}^{5} \frac{3y_i+1}{\frac{4}{5}} = \frac{5}{4}\sum_{i=1}^{5}(3y_i+1)$$

$$= \frac{5}{4} \times (3+5) = 10.$$

This completes the proof.

⬤ Arrange 1 650 students in 22 rows by 75 columns. It is known that for any two columns, the number of occasions that two students in the same row are of the same sex does not exceed 11. Prove that the number of boy students does not exceed 928. (posed by Feng Zhigang)

Solution Let a_i be the number of boy students in the ith row, then the number of girl students in this row is $75 - a_i$. By the given condition, we have $\sum_{i=1}^{22} (C_{a_i}^2 + C_{75-a_i}^2) \leqslant 11 \times C_{75}^2$. That is, $\sum_{i=1}^{22} (a_i^2 - 75a_i) \leqslant -30\ 525$, implying $\sum_{i=1}^{22} (2a_i - 75)^2 \leqslant 1\ 650$. Using Cauchy's Inequality, we have

$$\left[\sum_{i=1}^{22}(2a_i-75)\right]^2 \leqslant 22\sum_{i=1}^{22}(2a_i-75)^2 \leqslant 36\ 300.$$

Then $\sum_{i=1}^{22} (2a_i - 75) < 191$, and $\sum_{i=1}^{22} a_i < \dfrac{191 + 1\,650}{2} < 921$. That means the number of boy students does not exceed 928.

Remark This problem was derived from a more general one: Given a rectangular matrix with n rows and m colums and consisting of elements $+1$ and -1, if for any two colums, the sum of products of two elements in the same row is less than or equal to 0, find an upper bound of the number of elements $+1$ in the matrix. In our problem, $n = 22$ as there are 22 teams attending the competition, and there is an upper bound less than 928 when $m = 75$ as the competition date is September 28.

2004 (Yinchuan, Ningxia)

The 4th (2004) China Western Mathematical Olympiad was held on September 25 – 30, 2003 in Yinchuan, Ningxia, China, and was hosted by Ningxia Mathematical Society and Ningxia Changqing Yinchuan senior high school.

The Competition Committee consisted of the following: Xiong Bin, Wang Haiming, Xu Wanyi, Liu Shixiong, Wang Jianwei, Zhang Zhengjie, Wu Weichao, Feng Zhigang and Feng Yuefeng.

First Day
8:00 – 12:00 September 27, 2004

1 Find all integers n, such that $n^4 + 6n^3 + 11n^2 + 3n + 31$ is a perfect square. (posed by Xu Wanyi)

Solution Suppose $A = n^4 + 6n^3 + 11n^2 + 3n + 31$ is a perfect square, it means that $A = (n^2 + 3n + 1)^2 - 3(n - 10)$ is a perfect square.

If $n > 10$, then $A < (n^2 + 3n + 1)^2$, thus $A \leqslant (n^2 + 3n)^2$.

Therefore

$$(n^2 + 3n + 1)^2 - (n^2 + 3n)^2 \leqslant 3n - 30,$$

or

$$2n^2 + 3n + 31 \leqslant 0,$$

which is impossible.

If $n = 10$, then $A = (10^2 + 3 \times 10 + 1)^2 = 131^2$ is a perfect square.

If $n < 10$, then $A > (n^2 + 3n + 1)^2$.

Case I $n \leqslant -3$ or $0 \leqslant n < 10$. Then $n^2 + 3n \geqslant 0$. Thus

$$A \geqslant (n^2 + 3n + 2)^2.$$

That is,

$$2n^2 + 9n - 27 \leqslant 0,$$

or

$$-7 < \frac{-3(\sqrt{33} + 3)}{4} \leqslant n \leqslant \frac{3(\sqrt{33} - 3)}{4} < 3.$$

Therefore, $n = -6, -5, -4, -3, 0, 1, 2$. For these values of n, the corresponding values of A are 409, 166, 67, 40, 31, 52, 145. All of them are not perfect squares.

Case II $n = -2, -1$. Then $A = 37, 34$ respectively, none of them is a perfect square.

Hence, only when $n = 10$, A is a perfect square.

🔴 Let $ABCD$ be a convex quadrilateral, I_1 and I_2 be the incenters of $\triangle ABC$ and $\triangle DBC$ respectively. The line $I_1 I_2$ intersects the lines AB and DC at the points E and F respectively. Suppose lines AB and DC intersect at P, and $PE = PF$. Prove that the points A, B, C, D are concyclic. (posed by Liu Shixiong)

Solution   Draw line segments $I_1 B$, $I_1 C$, $I_2 B$ and $I_2 C$, as shown in the figure.

Since $PE = PF$, we have $\angle PEF = \angle PFE$.

But

$$\angle PEF = \angle I_2 I_1 B - \angle EBI_1 = \angle I_2 I_1 B - \angle I_1 BC,$$

and

$$\angle PFE = \angle I_1 I_2 C - \angle FCI_2 = \angle I_1 I_2 C - \angle I_2 CB.$$

Therefore, $\angle I_2 I_1 B - \angle I_1 BC = \angle I_1 I_2 C - \angle I_2 CB$, or

$$\angle I_2 I_1 B + \angle I_2 CB = \angle I_1 I_2 C + \angle I_1 BC.$$

On the other hand, $\angle I_2 I_1 B +$
$\angle I_2 CB + \angle I_1 I_2 C + \angle I_1 BC = 2\pi$. Thus
$\angle I_2 I_1 B + \angle I_2 CB = \pi$, and points I_1,
I_2, C, B are concyclic.

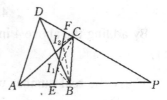

Since $\angle BI_1 C = \angle BI_2 C$, we have

$$\angle I_1 BC + \angle I_1 CB = \angle I_2 BC + \angle I_2 CB,$$

and

$$\angle ABC + \angle ACB = \angle DBC + \angle DCB.$$

Hence, $\angle BAC = \angle BDC$, and the points A, B, C, D are concyclic.

🔴 Find all real numbers k, such that the inequality

$$a^3 + b^3 + c^3 + d^3 + 1 \geqslant k(a+b+c+d)$$

holds for any a, b, c, $d \in [-1, +\infty)$. (posed by Xu Wanyi)

Solution　If $a = b = c = d = -1$, then $-3 \geqslant k \cdot (-4)$. Hence $k \geqslant \dfrac{3}{4}$.

If $a = b = c = d = \dfrac{1}{2}$, then $4 \cdot \dfrac{1}{8} + 1 \geqslant k \cdot \left(4 \cdot \dfrac{1}{2}\right)$. Thus $k \leqslant \dfrac{3}{4}$,

and so $k = \dfrac{3}{4}$.

Now we want to prove that the inequality

$$a^3 + b^3 + c^3 + d^3 + 1 \geqslant \frac{3}{4}(a+b+c+d) \tag{1}$$

holds for any a, b, c, $d \in [-1, +\infty)$.

At first, we prove that $4x^3 + 1 \geqslant 3x$, $x \in [-1, +\infty)$.

In fact, from $(x+1)(2x-1)^2 \geqslant 0$, we have $4x^3 + 1 \geqslant 3x$, $x \in [-1, +\infty)$. Therefore

$$4a^3 + 1 \geqslant 3a,$$

$$4b^3 + 1 \geqslant 3b,$$

$$4c^3 + 1 \geqslant 3c,$$

$$4d^3 + 1 \geqslant 3d.$$

By adding the above 4 inequalities together, we get the inequality (1).

Thus, the real number we want to find is $k = \dfrac{3}{4}$.

Remark To prove an inequality of the above type, it is natural that we prove an in equality for each variable separately and then add them together.

Let $n \in \mathbf{N}$ (the set of positive integers), and $d(n)$ be the number of positive divisors of n. Next, $\varphi(n)$ denotes the number of integers in the closed interval $[1, n]$ which are co-prime with n.

Find all non-negative integers c, such that there exists $n \in \mathbf{N}$ satisfying

$$d(n) + \varphi(n) = n + c.$$

For such c, find all n satisfying the above equation. (posed by Feng Zhigang)

Solution We denote the set of positive divisors of n by A, and the set of integers in the closed interval $[1, n]$ which are co-prime with n by B. Since there is only one number $1 \in A \cap B$ among $1, 2, \cdots, n$, we get $d(n) + \varphi(n) \leqslant n + 1$. Thus $c = 0$ or 1.

(1) If $c = 0$, then $d(n) + \varphi(n) = n$ implies that there exists only one number among $1, 2, \cdots, n$ which is not contained in $A \cup B$. If n is even, and $n > 8$, then all of $n - 2$ and $n - 4$ are not contained in $A \cup$

B. In this case, *n* does not satisfy the equation. If *n* is odd, then when *n* is a prime or 1, $d(n) + \varphi(n) = n + 1$ (which will be discussed in case (2) below). When *n* is a composite number, we can write $n = pq$, where *p* and *q* are odd and $1 < p \leqslant q$. If $q \geqslant 5$, then $2p$ and $4p$ are not contained in $A \cup B$, so *n* does not satisfy the equation.

From the above discussion, we find that

$$d(n) + \varphi(n) = n$$

when either $n \leqslant 8$ and *n* is even, or $n \leqslant 9$ and *n* is an odd composite number.

By direct verification of all the solutions of the above equation, *n* can only be 6, 8 and 9.

(2) If $c = 1$, then $d(n) + \varphi(n) = n + 1$ implies that among 1, 2, $\cdots$, *n* every number belongs to $A \cup B$. It is easy to find that this time $n = 1$ and primes can satisfy the required condition. For the case when *n* is even (not prime), by the same argument as above we have $n \leqslant 4$ (consider $n - 2$). If *n* is an odd composite number, write $n = pq$, where *p* and *q* are odd and $3 \leqslant p \leqslant q$, then $2p \notin A \cup B$, a contradiction. By direct calculation we find that only $n = 4$ satisfies the condition.

Therefore, the solutions of the equation $d(n) + \varphi(n) = n + 1$ are $n = 1$, 4 or primes.

Remark $A \cap B = \{1\}$ is clear, thus $c \leqslant 1$. When *n* is a prime, it is easy to see that $d(n) + \varphi(n) = n + 1$. When *n* is a composite number, for any composite number *n* which is large enough it is easy to find sufficiently many composite numbers less than *n* such that they are neither divisors of *n* nor coprime with *n*. In this case $d(n) + \varphi(n) \neq n + c$ ($c = 0$ or 1). The general case of this problem is to remove the restriction of *c* being non-negative.

Second Day

8:00 – 12:00 September 28, 2005

⑤ The sequence $\{a_n\}$ satisfies $a_1 = a_2 = 1$ and

$$a_{n+2} = \frac{1}{a_{n+1}} + a_n, \quad n = 1, 2, \cdots.$$

Find $a_{2\,004}$. (posed by Wu Weichao)

Solution   According to the assumption we have

$$a_{n+2}a_{n+1} - a_{n+1}a_n = 1.$$

Thus, $\{a_{n+1}a_n\}$ is an arithmetic progression with first term 1 and common difference 1. Hence

$$a_{n+1}a_n = n, \quad n = 1, 2, \cdots.$$

So $a_{n+2} = \dfrac{n+1}{a_{n+1}} = \dfrac{n+1}{\dfrac{n}{a_n}} = \dfrac{n+1}{n}a_n$, $n = 1, 2, \cdots$. Consequently,

$$a_{2\,004} = \frac{2\,003}{2\,002}a_{2\,002} = \frac{2\,003}{2\,002} \cdot \frac{2\,001}{2\,000}a_{2\,000}$$

$$= \cdots = \frac{2\,003}{2\,002} \cdot \frac{2\,001}{2\,000} \cdot \cdots \cdot \frac{3}{2}a_2$$

$$= \frac{3 \cdot 5 \cdot \cdots \cdot 2\,003}{2 \cdot 4 \cdot \cdots \cdot 2\,002}.$$

⑤ All the grids of an $m \times n$ chessboard ($m \geqslant 3$, $n \geqslant 3$) are colored either red or blue. Two adjacent grids (with a common side) are called a *good couple* if they are of different colors. Suppose that there are S good couples, explain how to determine whether S is odd or even. Does it depend on certain specific color grids? (Reasoning is required.) (posed by Feng Yuefeng)

Solution I   Classify all grids into three parts: the grids at the four corners, the grids along the borderlines (not including four corners), and the other grids. Fill all red grids with label number 1, all blue grids with label number -1. Denote the label numbers filled in the grids in the first part by a, b, c and d, in the second part by x_1, x_2, $\cdots$, $x_{2m+2n-8}$, and in the third part by y_1, y_2, $\cdots$, $y_{(m-2)(n-2)}$. For any two adjacent grids we write a label number which is the product of two label numbers of the two grids on their common edge. Let H be the product of all label numbers on common edges.

There are 2 adjacent grids for every grid in the first part, thus its label number appears twice in H. There are 3 adjacent grids for every grid in the second part, thus its label number appears three times in H. There are 4 adjacent grids for every grid in the third part, thus its label number appears four times in H. Therefore,

$$H = (abcd)^2 (x_1 x_2 \cdots x_{2m+2n-8})^3 (y_1 y_2 \cdots y_{(m-2)(n-2)})^4$$

$$= (x_1 x_2 \cdots x_{2m+2n-8})^3.$$

If $x_1 x_2 \cdots x_{2m+2n-8} = 1$, then $H = 1$, and in this case there are even good couples. If $x_1 x_2 \cdots x_{2m+2n-8} = -1$, then $H = -1$, and in this case there are odd good couples. It shows that whether S is even or odd is determined by colors of the grids in the second part. Moreover, when there are odd blue grids among the grids in the second part, S is odd. Otherwise S is even.

Solution Ⅱ   Classify all grids into three parts: the grids at the four corners, the grids along the borderlines (not containing four corners), and the other grids.

If all grids are red, then $S = 0$, which is even. If there are blue grids we pick any one of them, say A, and change A into a red one. We call this changing a transformation.

(1) A is a grid in the first part. Suppose that there are k red grids and $2 - k$ blue grids among A's two adjacent grids. After changing A into a red one, the number of good couples increases by $2 - k - k = 2 - 2k$. It follows that the parity even or odd of S is unchanged.

(2) A is a grid in the second part. Suppose that there are p red grids and $3 - p$ blue grids among A's three adjacent grids. After changing A into a red one, the number of good couples increases by $3 - p - p = 3 - 2p$. It follows that the parity of S is changed.

(3) A is a grid in the third part. Suppose that there are q red

grids and $4 - q$ blue grids among A's four adjacent grids. After changing A into a red one, the number of good couples increases by $4 - q - q = 4 - 2q$. It follows that the parity of S is unchanged.

If there are still blue grids on the chessboard after above transformation, we continue doing the transformation over and over again until no blue grid left on the chessboard. Now S is changed into 0.

Clearly, S changes its parity odd times if there are odd blue grids among the second part of grids. Similarly, S changes its parity even times if there are even blue grids among the second part of grids. It implies that the parity of S is determined by the coloring of the second part of grids. When there exist odd blue grids among the second part of grids, S is odd. When there exist even blue grids among the second part of grids, S is even.

Remark In Solution I the method of evaluation is used, and in Solution II the method of transformation is used. These two methods are used quite often in solving this kind of problems.

 Let l be the perimeter of an acute triangle $\triangle ABC$ which is not equilateral, P a variable point inside $\triangle ABC$, and D, E and F be projections of P on BC, CA and AB respectively.

Prove that

$$2(AF + BD + CE) = l,$$

if and only if P is collinear with the incenter and circumcenter of $\triangle ABC$. (posed by Xiong Bin)

Solution Denote the lengths of three sides of $\triangle ABC$ by $BC = a$, $CA = b$ and $AB = c$ respectively. No loss of generality, we can suppose $b \neq c$. We choose a rectangular coordinate system (see the figure), then we have $A(m, n)$, $B(0, 0)$, $C(a, 0)$ and $P(x, y)$.

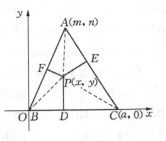

Since $AF^2 - BF^2 = AP^2 - BP^2$, it follows that

$$AF^2 - (c-AF)^2 = AP^2 - BP^2.$$

Therefore,

$$2c \cdot AF - c^2 = (x-m)^2 + (y-n)^2 - x^2 - y^2,$$

and

$$AF = \frac{m^2 + n^2 - 2mx - 2ny}{2c} + \frac{c}{2}.$$

On the other hand, from

$$CE^2 - AE^2 = PC^2 - AP^2,$$

we have

$$CE^2 - (b-CE)^2 = PC^2 - AP^2.$$

Thus

$$2b \cdot CE - b^2 = (x-a)^2 + y^2 - (x-m)^2 - (y-n)^2,$$

$$CE = \frac{2mx + 2ny - m^2 - n^2 - 2ax + a^2}{2b} + \frac{b}{2}.$$

Since $AF + BD + CE = \dfrac{l}{2}$, we get

$$\frac{m^2 + n^2 - 2mx - 2ny}{2c} + \frac{c}{2} + x + \frac{2mx + 2ny - m^2 - n^2 - 2ax + a^2}{2b}$$

$$+ \frac{b}{2} = \frac{l}{2},$$

that is,

$$\left(\frac{m}{b} - \frac{a}{b} - \frac{m}{c} + 1\right)x + \left(\frac{n}{b} - \frac{n}{c}\right)y + \frac{a^2}{2b} + \frac{b+c}{2}$$

$$+ \frac{m^2 + n^2}{2}\left(\frac{1}{c} - \frac{1}{b}\right) - \frac{l}{2} = 0.$$

Since $b \neq c$ and $n \neq 0$, point P is on a fixed straight line. Since the condition $2(AF + BD + CE) = l$ is satisfied for both incenter and circumcenter, we complete the proof.

Remark   The method we used above is analytic. Another solution using a purely geometric method is as follows: construct projections of the incenter, circumcenter of $\triangle ABC$ and point P on the three sides of $\triangle ABC$ respectively, we can get the result by means of proportional segments (see the figure).

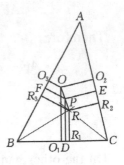

Suppose that a, b, c are positive real numbers, prove that

$$1 < \frac{a}{\sqrt{a^2+b^2}} + \frac{b}{\sqrt{b^2+c^2}} + \frac{c}{\sqrt{c^2+a^2}} \leqslant \frac{3\sqrt{2}}{2}.$$

(posed by Wang Jianwei)

Solution   Set $x = \dfrac{b^2}{a^2}$, $y = \dfrac{c^2}{b^2}$, $z = \dfrac{a^2}{c^2}$, then x, y, $z \in \mathbf{R}^+$ and $xyz = 1$. It suffices to prove that

$$1 < \frac{1}{\sqrt{1+x}} + \frac{1}{\sqrt{1+y}} + \frac{1}{\sqrt{1+z}} \leqslant \frac{3\sqrt{2}}{2}.$$

Without loss of generality, we assume that $x \leqslant y \leqslant z$. Set $A = xy$, we have $z = \dfrac{1}{A}$, $A \leqslant 1$. Thus

$$\frac{1}{\sqrt{1+x}} + \frac{1}{\sqrt{1+y}} + \frac{1}{\sqrt{1+z}} > \frac{1}{\sqrt{1+x}} + \frac{1}{\sqrt{1+\dfrac{1}{x}}} = \frac{1+\sqrt{x}}{\sqrt{1+x}} > 1.$$

Let $u = \dfrac{1}{\sqrt{1+A+x+\dfrac{A}{x}}}$, then $u \in \left(0, \dfrac{1}{1+\sqrt{A}}\right]$, and $x = \sqrt{A}$ if and only if $u = \dfrac{1}{1+\sqrt{A}}$. Hence

$$\left(\frac{1}{\sqrt{1+x}} + \frac{1}{\sqrt{1+y}}\right)^2 = \left[\frac{1}{\sqrt{1+x}} + \frac{1}{\sqrt{1+\dfrac{A}{x}}}\right]^2$$

$$= \frac{1}{1+x} + \frac{1}{1+\frac{A}{x}} + \frac{2}{\sqrt{1+A+x+\frac{A}{x}}}$$

$$= \frac{2+x+\frac{A}{x}}{1+A+x+\frac{A}{x}} + \frac{2}{\sqrt{1+A+x+\frac{A}{x}}}$$

$$= 1 + (1-A)u^2 + 2u.$$

Set $f(u) = (1-A)u^2 + 2u + 1$, we see that $f(u)$ is an increasing function on $u \in \left(0, \dfrac{1}{1+\sqrt{A}}\right]$, which implies that

$$\frac{1}{\sqrt{1+x}} + \frac{1}{\sqrt{1+y}} \leqslant \sqrt{f\left(\frac{1}{1+\sqrt{A}}\right)} = \frac{2}{\sqrt{1+\sqrt{A}}}.$$

Now set $\sqrt{A} = v$, we get

$$\frac{1}{\sqrt{1+x}} + \frac{1}{\sqrt{1+y}} + \frac{1}{\sqrt{1+z}} \leqslant \frac{2}{\sqrt{1+\sqrt{A}}} + \frac{1}{\sqrt{1+\frac{1}{A}}}$$

$$= \frac{2}{\sqrt{1+v}} + \frac{\sqrt{2}v}{\sqrt{2(1+v^2)}} \leqslant \frac{2}{\sqrt{1+v}} + \frac{\sqrt{2}v}{1+v}$$

$$= \frac{2}{\sqrt{1+v}} + \sqrt{2} - \frac{\sqrt{2}}{1+v} = -\sqrt{2}\left[\frac{1}{\sqrt{1+v}} - \frac{\sqrt{2}}{2}\right]^2 + \frac{3\sqrt{2}}{2}$$

$$\leqslant \frac{3\sqrt{2}}{2}.$$

2005   (Chengdu, Sichuan)

The 5th (2005) China Western Mathematical Olympiad was held on

November 3 - 11, 2005 in Chengdu, Sichuan, China, and was hosted by Sichuan Mathematical Society and Chengdu No. 7 middle school.

The Competition Committee consisted of the following: Li Shenghong, Tang Xianjiang, Leng Gangsong, Li Weigu, Zhu Huawei, Weng Kaiqing, Tang Lihua and Bian hongping.

First Day
8:00 - 12:00 November 5, 2005

① Assume that $\alpha^{2\,005} + \beta^{2\,005}$ can be expressed as a polynomial in $\alpha + \beta$ and $\alpha\beta$. Find the sum of the coefficients of the polynomial. (posed by Zhu Huawei)

Solution I   In the expansion of $\alpha^k + \beta^k$, let $\alpha + \beta = 1$ and $\alpha\beta = 1$. We get the sum of coefficients $S_k = \alpha^k + \beta^k$. Since

$$(\alpha + \beta)(\alpha^{k-1} + \beta^{k-1})$$
$$= (\alpha^k + \beta^k) + \alpha\beta(\alpha^{k-2} + \beta^{k-2}),$$

we get $S_k = S_{k-1} - S_{k-2}.$

Thus $S_k = S_{k-6}$ and $\{S_k\}$ is a periodic sequence with period 6 and $S_{2\,005} = S_1 = 1$.

Solution II   Set $\alpha + \beta = 1$ and $\alpha\beta = 1$ in the expansion of $\alpha^k + \beta^k$. The sum of the coefficients is $S_k = \alpha^k + \beta^k$. Since α, β are solutions of the equation $x^2 - x + 1 = 0$, $\alpha = \cos\dfrac{\pi}{3} + i\sin\dfrac{\pi}{3}$, $\beta = \cos\dfrac{\pi}{3} - i\sin\dfrac{\pi}{3}$. Therefore

$$\alpha^k + \beta^k = \left(\cos\frac{\pi}{3} + i\sin\frac{\pi}{3}\right)^k + \left(\cos\frac{\pi}{3} - i\sin\frac{\pi}{3}\right)^k$$

$$= \left(\cos\frac{k\pi}{3} + i\sin\frac{k\pi}{3}\right) + \left(\cos\frac{k\pi}{3} - i\sin\frac{k\pi}{3}\right)$$

$$= 2\cos\frac{k\pi}{3}.$$

Let $k = 2\,005$, we have $S_k = 1$.

As shown in the diagram, PA, PB are two tangent lines of a circle from a point P outside the circle, and A, B are the contact points. PD is a secant line, and it intersects the circle at C and D. BF parallels PA and meets the lines AC, AD at E, F respectively. Prove that $BE = BF$.

(posed by Leng Gangrong)

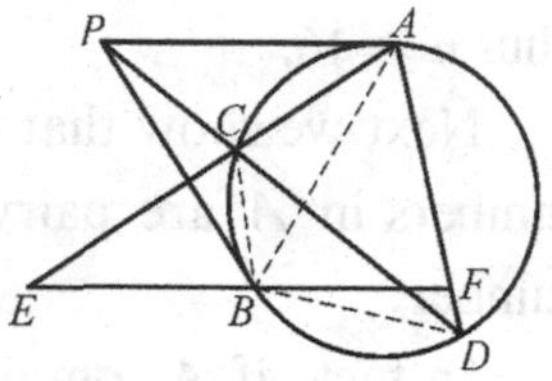

Proof Join BC, BA and BD, then $\angle ABC = \angle PAC = \angle E$. Thus $\triangle ABC \backsim \triangle AEB$ and $\dfrac{BC}{BE} = \dfrac{AC}{AB}$, that is,

$$BE = \frac{AB \cdot BC}{AC}. \qquad \qquad ①$$

Since $\angle ABF = \angle PAB = \angle ADB$, we have $\triangle ABF \backsim \triangle ADB$. Therefore $\dfrac{BF}{BD} = \dfrac{AB}{AD}$, that is,

$$BF = \frac{AB \cdot BD}{AD}. \qquad \qquad ②$$

On the other hand, since $\triangle PBC \backsim \triangle PDB$ and $\triangle PCA \backsim \triangle PAD$, we get

$$\frac{BC}{BD} = \frac{PC}{PB} \text{ and} \frac{AC}{AD} = \frac{PC}{PA}.$$

Since $PA = PB$, we have

$$\frac{BC}{BD} = \frac{AC}{AD}. \qquad \qquad ③$$

Thus $\dfrac{BC}{AC} = \dfrac{BD}{AD}$. By ①, ② and ③, we have $BE = BF$.

Let $S = \{1, 2, \cdots, 2\,005\}$. If there is at least one prime number in any subset of S consisting of n pairwise coprime numbers, find the minimum value of n. (posed by Tang Lihua)

Solution   First we prove $n \geqslant 16$. In fact, let

$$A_0 = \{1, 2^2, 3^2, 5^2, \cdots, 41^2, 43^2\},$$

where the members in A_0, other than 1, are the squares of prime numbers not greater than 43. Then $A_0 \subseteq S$, $|A_0| = 15$ and the numbers in A_0 are pairwise coprime but A_0 contains no prime number. Thus $n \geqslant 16$.

Next we show that for arbitrary $A \subseteq S$ with $n = |A| = 16$, if the numbers in A are pairwise coprime, then A must contain a prime number.

In fact, if A contains no prime number, denote $A = \{a_1, a_2, \cdots, a_{16}; a_1 < a_2 < \cdots < a_{16}\}$. Then there are two possibilities.

(1) If $1 \notin A$, then $a_1, a_2, \cdots, a_{16}$ are composite number. Since $(a_i, a_j) = 1 \ (1 \leqslant i < j \leqslant 16)$, the prime factors of a_i and a_j are pairwise distinct. Let p_i be the smallest prime factor of a_i. We may assume that $p_1 < p_2 < \cdots < p_{16}$, then

$$a_1 \geqslant p_1^2 \geqslant 2^2, \ a_2 \geqslant p_2^2 \geqslant 3^2, \cdots, a_{15} \geqslant p_{15}^2 \geqslant 47^2 > 2\,005,$$

it leads to a contradiction.

(2) If $1 \in A$, let $a_{16} = 1$, $a_1, a_2, \cdots, a_{15}$ are composite numbers. By the same assumption and argument of (1), we have $a_1 \geqslant p_1^2 \geqslant 2^2$, $a_2 \geqslant p_2^2 \geqslant 3^2, \cdots, a_{15} \geqslant p_{15}^2 \geqslant 47^2 > 2\,005$. Again, it leads to a contradiction.

From (1) and (2), A contains at least one prime number, i. e. when $n = |A| = 16$, the conclusion is true. Thus, the minimum number of n is 16.

It is given that real numbers $x_1, x_2, \cdots, x_n \ (n > 2)$ satisfy $\left| \sum_{i=1}^{n} x_i \right| > 1, |x_i| \leqslant 1 \ (i = 1, 2, \cdots, n)$. Prove that there exists a positive integer k such that $\left| \sum_{i=1}^{k} x_i - \sum_{i=k+1}^{n} x_i \right| \leqslant 1$. (posed by Leng Gangsong)

Proof Set $g(0) = -\sum_{i=1}^{n} x_i$, $g(k) = \sum_{i=1}^{k} x_i - \sum_{i=k+1}^{n} x_i (1 \leqslant k \leqslant n-1)$,

$g(n) = \sum_{i=1}^{n} x_i$.

Then
$$|g(1) - g(0)| = 2|x_1| \leqslant 2,$$

$$|g(k+1) - g(k)| = 2|x_{k+1}| \leqslant 2, \; k = 1, 2, \cdots, n-2,$$

$$|g(n) - g(n-1)| = 2|x_n| \leqslant 2.$$

So for each $0 \leqslant k \leqslant n-1$,

$$|g(k+1) - g(k)| \leqslant 2. \qquad\qquad ①$$

If the conclusion is not true, by the condition for each k, $0 \leqslant k \leqslant n$, we have

$$|g(k)| > 1. \qquad\qquad ②$$

If there is an i, $0 \leqslant i \leqslant n-1$, such that $g(i)g(i+1) < 0$, we may assume that $g(i) > 0$ and $g(i+1) < 0$. By ②, $g(i) > 1$ and $g(i+1) < -1$. Thus $|g(i+1) - g(i)| > 2$. This contradicts ①. Thus $g(0)$, $g(1)$, $\cdots$, $g(n)$ have the same sign. But $g(0) + g(n) = 0$. The contradiction implies that the conclusion is true.

Second Day
8:00 – 12:00 November 6, 2005

⑤ The circles O_1 and O_2 meet at points A and B. The line DC passes through O_1, intersects the circle O_1 at D and is a tangent to the circle O_2 at C. Also, CA is a tangent to the circle O_1 at A. The secant AE of the circle O_1 is perpendicular to DC. AF is perpendicular to and meets DE at F.

Prove that BD bisects the line segment AF. (posed by Bian Hongping)

Proof Let AE intersect DC at point H,

and AF intersect BD at point G. Join AB, BC, BH, BE, CE and GH. By symmetry, CE is also a tangent line of the circle O_1 and H is the midpoint of AE.

Since $\angle HCB = \angle BAC$ and $\angle BAC = \angle BEH$, we have $\angle HCB = \angle HEB$. Thus H, B, C, E lie on the same circle, and $\angle BHC = \angle BEC$.

From
$$\angle BEC = \angle BDE,$$

we get
$$\angle BHC = \angle BDE. \qquad \text{①}$$

Since $AF \perp DE$, we have

$$\angle AGB = \frac{\pi}{2} - \angle BDE. \qquad \text{②}$$

By $AE \perp DC$,

$$\angle AHB = \frac{\pi}{2} - \angle BHC. \qquad \text{③}$$

By ①, ② and ③, $\angle AGB = \angle AHB$. Therefore A, G, H, B lie on the same circle, and $\angle AHG = \angle ABG = \angle AED$. Thus $GH /\!/ DE$. Since H is the midpoint of AE, G is the midpoint of AF.

In an isosceles right angled triangle $\triangle ABC$, $CA = CB = 1$, and P is an arbitrary point on the perimeter of $\triangle ABC$. Find the maximum value of $PA \cdot PB \cdot PC$. (posed by Li Weigu)

Solution

(1) In the first diagram, if $P \in AC$, we have $PA \cdot PC \leqslant \frac{1}{4}$ and $PB \leqslant \sqrt{2}$. Thus $PA \cdot$

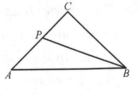

$PB \cdot PC \leqslant \frac{\sqrt{2}}{4}$. The equality is not valid, since the two equality signs cannot be valid at the same time. Therefore $PA \cdot PB \cdot PC < \frac{\sqrt{2}}{4}$.

(2) In the second diagram, if $P \in AB$, write $AP = x \in [0, \sqrt{2}]$,

then

$$f(x) = PA^2 \cdot PB^2 \cdot PC^2$$

$$= x^2(\sqrt{2} - x)^2(1 + x^2 - \sqrt{2}x).$$

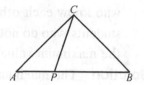

Let $t = x(\sqrt{2} - x)$, then $t \in \left[0, \frac{1}{2}\right]$ and $f(x) = g(t) = t^2(1 - t)$.

Note that $g'(t) = 2t - 3t^2 = t(2 - 3t)$. Thus $g(t)$ is increasing on $\left[0, \frac{2}{3}\right]$ and $f(x) \leqslant g\left(\frac{1}{2}\right) = \frac{1}{8}$. Therefore $PA \cdot PB \cdot PC \leqslant \frac{1}{2\sqrt{2}} = \frac{\sqrt{2}}{4}$. The equality is valid if and only if $t = \frac{1}{2}$ and $x = \frac{\sqrt{2}}{2}$. So P is the midpoint of AB.

Given real numbers a, b, c, satisfying $a + b + c = 1$, prove that
$$10(a^3 + b^3 + c^3) - 9(a^5 + b^5 + c^5) \geqslant 1. \quad \text{(posed by Li Shenghong)}$$

Solution   Since $\sum a^3 = 1 - 3\prod(a + b)$, $\sum a^5 = 1 - 5\prod(a + b)\left[\sum a^2 + \sum ab\right]$,

therefore, the original inequality holds

$$\Leftrightarrow 10[1 - 3\prod(a + b)] - 9\left[1 - 5\prod(a + b)(\sum a^2 + \sum ab)\right] \geqslant 1$$

$$\Leftrightarrow 45\prod(a + b)(\sum a^2 + \sum ab) \geqslant 30\prod(a + b)$$

$$\Leftrightarrow 3(\sum a^2 + \sum ab) \geqslant 2 = 2(\sum a)^2 = 2(\sum a^2 + 2\sum ab)$$

$$\Leftrightarrow \sum a^2 \geqslant \sum ab.$$

From $a^2 + b^2 \geqslant 2ab$, $b^2 + c^2 \geqslant 2bc$ and $c^2 + a^2 \geqslant 2ac$, we have $2\sum a^2 \geqslant 2\sum ab$, i. e., $\sum a^2 \geqslant \sum ab$. Therefore the original inequality holds.

There are n new students. Suppose that there are two students who know each other in every three students and there are two students who do not know each other in every four students. Find the maximum value of n. (posed by Tang Lihua)

Solution The maximum value of n is 8.

When $n = 8$, the example shown in the diagram satisfies the requirements, where A_1, A_2, $\cdots$, A_8 represent 8 students. The line segment between A_i and A_j means A_i and A_j know each other.

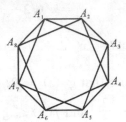

Next, if n students satisfy the conditions, we want to show that $n \leqslant 8$. To do this, we first prove that the following two cases are impossible.

(1) If someone A knows at least 6 persons, denoted by B_1, B_2, $\cdots$, B_6. By Ramsey's theorem, there exist 3 persons among them who do not know each other. This contradicts that there are two who know each other in every three students, or that there exist 3 people they know each other. A and the three persons form a group of four persons such that every two of them know each other. A contradiction.

(2) If some one A knows at most $n - 5$ persons, then in the remaining there are at least 4 persons, none of whom knows A. Thus every two of the four persons know each other, a contradiction.

When $n \geqslant 10$, one of (1) and (2) must occur. So such n does not satisfy the requirements.

If $n = 9$, in order to avoid (1) and (2), each person knows exactly 5 other persons. Thus the number of pairs knowing each other is $\dfrac{9 \times 5}{2} \notin \mathbf{N}$. This contradiction implies $n \leqslant 8$. Thus the maximum value of n is 8.

International Mathematical Olympiad

The international Mathematical Olympiad, founded in 1959, is one of the most competitive and highly intellectual activities in the world. Till now, there are more than 90 countries and areas that take part in it. IMO, hosted by each participating country in turn, is held in mid-July every year. The competition lasts for 2 days, and there are 3 problems to be completed within 4.5 hours each day. Each question is 7 marks which total up to 42 marks. The full score for a team is 252 marks. About half of the participants will be awarded a medal, where 1/12 will be awarded a gold medal. The numbers of gold, silver, and bronze medals awarded are in the ratio of 1 : 2 : 3 approximately. In the case when a participant provides a better solution than the official answer, a special award is given.

All participating countries are required to send a delegation consisting of a leader, a deputy leader and 6 contestants. The problems are contributed by the participating countries and are later selected carefully by the host country for submission to the international jury set up by the host country. The host country does not provide any question. Then the problems are translated in working languages and the team leaders will translate them into there own languages.

The answer scripts of each participating team will be marked by the team leader and the deputy leader. The team leader will later present them to the coordinators for assessment. If there is any dispute, the matter will be settled by the jury. The jury is formed by the various team leaders and an appointed chairman by the host country. The jury is responsible for deciding the final 6 problems, finalizing the marking standard, ensuring the accuracy of the translation of the problems, standardizing replies to written queries raised by participants during the competition, synchronizing differences in marking between the leaders and the coordinators and deciding on the cut-off points for the medals depending on the contestants' results as the difficulties of problems each year are different.

2003 (Tokyo, Japan)

The 44th IMO (International Mathematical Olympiad) was hosted by Japan in Tokyo during July 7 – 19 in 2003

The leader of Chinese IMO 2003 team was Prof. Li Shenghong who was from Zhejiang University and the deputy leader was Feng Zhigang who was from Shanghai High School. In IMO 2003, China came second among the nations, with two golds, one silver. Here are the results:

Fu Yunhao	the High School Attached to Tsinghua University	42 points	gold medal
Wang Wei	the High School Attached to Hunan Normal University	37 points	gold medal
Xiang Zhen	the First High School of Changsha, Hunan	36 points	gold medal
Fang Jiacong	the Affiliated High School of South China Normal University	35 points	gold medal
Wan Xin	Sichuan Pengzhou Middle School	33 points	gold medal
Zhou You	No. 3 High School of WISCO (a second-year student)	28 points	silver medal

First Day
9:00 - 13:30 July 13, 2003

Let A be a subset of the set $S = \{1, 2, \cdots, 1\,000\,000\}$ containing exactly 101 elements. Prove that there exist numbers t_1, t_2, $\cdots$, t_{100} in S such that the sets

$$A_j = \{x + t_j \mid x \in A\} \text{ for } j = 1, 2, \cdots, 100$$

are pairwise disjoint.

Solution Consider the set $D = \{x - y \mid x, y \in A\}$. There are at most $101 \times 100 + 1 = 10\,101$ elements in D. Two sets A_i and A_j have nonempty intersection if and only if $t_i - t_j$ is in D. So we need to choose the 100 elements in such a way that the difference for any two elements is not in D.

Now select these elements by induction. Choose one element arbitrarily from S. Assume that k elements, $k \leqslant 99$, are already chosen. An element x in S that is already chosen prevents us from selecting any element from the set $x + D$, where $x + D = \{x + y \mid y \in D\}$. Thus after k elements are chosen, at most $10\,101k \leqslant 999\,999$

elements in S are forbidden. Hence we can select one more element.

Remark   The size $|S| = 10^6$ is unnecessarily large. The following statement is true:

If A is a k-element subset of $S = \{1, 2, \cdots, n\}$ and m is a positive integer such that $n > (m-1)\left(\binom{k}{2}+1\right)$, then there exist $t_1, \cdots, t_m \in S$ such that the sets $A_j = \{x+t_j \mid x \in A\}$ for $j = 1, \cdots, m$ are pairwise disjoint.

⟐ Determine all pairs of positive integers (a, b) such that

$$\frac{a^2}{2ab^2 - b^3 + 1}$$

is a positive integer.

Solution I   Let (a, b) be a pair of positive integers satisfying the condition. Since $k = \dfrac{a^2}{2ab^2 - b^3 + 1} > 0$, we have $2ab^2 - b^3 + 1 > 0$ or $a > \dfrac{b}{2} - \dfrac{1}{2b^2}$, and hence $a \geqslant \dfrac{b}{2}$. Using this, we infer from $k \geqslant 1$, or $a^2 \geqslant b^2(2a - b) + 1$, that $a^2 > b^2(2a - b) \geqslant 0$. Hence

$$a > b \text{ or } 2a = b. \tag{①}$$

Now consider the two solutions a_1, a_2 of the equation

$$a^2 - 2kb^2 a + k(b^3 - 1) = 0 \tag{②}$$

for any fixed positive integers k and b, and assume that one of them is an integer. Then the other is also an integer because $a_1 + a_2 = 2kb^2$. We may assume that $a_1 \geqslant a_2$, and we have $a_1 \geqslant kb^2 > 0$. Furthermore, since $a_1 a_2 = k(b^3 - 1)$, we get

$$0 \leqslant a_2 = \frac{k(b^3 - 1)}{a_1} \leqslant \frac{k(b^3 - 1)}{kb^2} < b.$$

Together with ①, we conclude that $a_2 = 0$ or $a_2 = \dfrac{b}{2}$ (in the latter case b must be even).

If $a_2 = 0$, then $b^3 - 1 = 0$, and hence $a_1 = 2k$ and $b = 1$.

If $a_2 = \dfrac{b}{2}$, then $k = \dfrac{b^2}{4}$ and $a_1 = \dfrac{b^4}{2} - \dfrac{b}{2}$.

Therefore the only possibilities are

$$(a, b) = (2l, 1), \ (l, 2l) \text{ or } (8l^4 - l, 2l)$$

for some positive integer l. All of these pairs satisfy the given condition.

Solution Ⅱ If $b = 1$, it follows from the given condition that a must be even.

Let $\dfrac{a^2}{2ab^2 - b^3 + 1} = k$. If $b > 1$, then there are two solutions to the equation ② and one of them is a positive integer. Thus the discriminant Δ of the equation ② is a perfect square, that is $\Delta = 4k^2b^4 - 4k(b^3 - 1)$ is a perfect square.

Note that, if $b \geqslant 2$, we have

$$(2kb^2 - b - 1)^2 < \Delta < (2kb^2 - b + 1)^2. \qquad ③$$

The proof is given as follows,

$$\Delta - (2kb^2 - b - 1)^2 = 4kb^2 - b^2 - 2b + 4k - 1$$
$$= (4k - 1)(b^2 + 1) - 2b$$
$$\geqslant 2(4k - 1)b - 2b > 0,$$
$$(2kb^2 - b + 1)^2 - \Delta = 4kb^2 - 4k - (b - 1)^2$$
$$= 4k(b^2 - 1) - (b - 1)^2$$
$$> (4k - 1)(b^2 - 1) > 0,$$

this completes the proof of ③.

Since Δ is a perfect square, it follows from ③ that

$$\Delta = 4k^2b^4 - 4k(b^3 - 1) = (2kb^2 - b)^2.$$

Then $4k = b^2$, and hence b must be even. Let $b = 2l$. We have $k = l^2$. Together with ②, we have $a = l$ or $8l^4 - l$.

Therefore the only possibilities are

$$(a, b) = (2l, 1), (l, 2l) \text{ or } (8l^4 - l, 2l)$$

for some positive integer l. All of these pairs satisfy the given condition.

A convex hexagon is given in which any two opposite sides have the following property: the distance between their midpoints is $\dfrac{\sqrt{3}}{2}$ times the sum of their lengths. Prove that all the angles of the hexagon are equal.

(A convex hexagon $ABCDEF$ has three pairs of opposite sides: AB and DE, BC and EF, CD and FA.)

Proof I   We first prove the following lemma.

Lemma *Consider a triangle PQR with $\angle QPR \geqslant 60°$. Let L be the midpoint of QR. Then $PL \leqslant \dfrac{\sqrt{3}}{2} QR$. The equality holds if and only if the triangle PQR is equilateral.*

Proof of the lemma

Let S be the point such that the triangle QRS is equilateral, where the points P and S lie in the same half-plane bounded by the line QR.

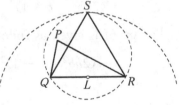

Then the point P lies inside the circumcircle (including the circumference of the circle) of the triangle QRS, which lies inside the circle with center L and radius $\dfrac{\sqrt{3}}{2} QR$.

This completes the proof of the lemma.

The main diagonals of a convex hexagon form a triangle though the

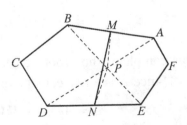

triangle can be degenerated. Thus we may choose two of these three diagonals that form an angle greater than or equal to 60°. Without loss of generality, we may assume that the diagonals AD and BE of the given hexagon $ABCDEF$ satisfy $\angle APB \geqslant 60°$, where P is the intersection of these two diagonals. Then, using the lemma, we obtain

$$MN = \frac{\sqrt{3}}{2}(AB + DE) \geqslant PM + PN \geqslant MN,$$

where M and N are the midpoints of AB and DE respectively. Thus it follows from the lemma that the triangles ABP and DEP are equilateral.

Therefore the diagonal CF forms an angle greater than or equal to 60° with one of the diagonals AD and BE. Without loss of generality, we may assume that $\angle AQF \geqslant 60°$, where Q is the intersection of AD and CF. Arguing in the same way as above, we infer that the triangles AQF and CQD are equilateral. This implies that $\angle BRC = 60°$, where R is the intersection of BE and CF. Using the same argument as above for the third time, we obtain that the triangles BCR and EFR are equilateral. This completes the proof.

Proof II　　Let $ABCDEF$ be the given hexagon and let $a = \overrightarrow{AB}$, $b = \overrightarrow{BC}$, $\cdots$, $f = \overrightarrow{FA}$.

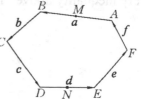

Let M and N be the midpoints of the sides AB and DE respectively. We have

$$\overrightarrow{MN} = \frac{1}{2}a + b + c + \frac{1}{2}d \text{ and } \overrightarrow{MN} = -\frac{1}{2}a - f - e - \frac{1}{2}d.$$

Thus we obtain

$$\overrightarrow{MN} = \frac{1}{2}(b + c - e - f). \qquad ①$$

From the given property, we have

$$\overrightarrow{MN} = \frac{\sqrt{3}}{2}(|a| + |d|) \geqslant \frac{\sqrt{3}}{2}|a - d|. \qquad ②$$

Set $x = a - d$, $y = c - f$, $z = e - b$. From ① and ②, we obtain

$$|y - z| \geqslant \sqrt{3}|x|. \qquad ③$$

Similarly, we see that

$$|z - x| \geqslant \sqrt{3}|y|, \qquad ④$$

$$|x - y| \geqslant \sqrt{3}|z|. \qquad ⑤$$

Note that

$$③ \Leftrightarrow |y|^2 - 2y \cdot z + |z|^2 \geqslant 3|x|^2,$$
$$④ \Leftrightarrow |z|^2 - 2z \cdot x + |x|^2 \geqslant 3|y|^2,$$
$$⑤ \Leftrightarrow |x|^2 - 2x \cdot y + |y|^2 \geqslant 3|z|^2.$$

By adding up the last three inequalities, we obtain

$$-|x|^2 - |y|^2 - |z|^2 - 2y \cdot z - 2z \cdot x - 2x \cdot y \geqslant 0,$$

or $-|x + y + z|^2 \geqslant 0$. Thus $x + y + z = 0$ and the equality holds in every inequality above. Hence we conclude that

$$x + y + z = 0,$$

$$|y - z| = \sqrt{3}|x|, \ a \ /\!/ \ d \ /\!/ \ x,$$

$$|z - x| = \sqrt{3}|y|, \ c \ /\!/ \ f \ /\!/ \ y,$$

$$|x - y| = \sqrt{3}|z|, \ e \ /\!/ \ b \ /\!/ \ z.$$

Suppose that PQR is the triangle such that $\overrightarrow{PQ} = x$, $\overrightarrow{QR} = y$, $\overrightarrow{RP} = z$. We may assume $\angle QPR \geqslant 60°$, without loss of generality. Let L be the midpoint of QR. Then $PL = \frac{1}{2}|z - x| = \frac{\sqrt{3}}{2}|y| = \frac{\sqrt{3}}{2}QR$. It follows from the lemma in Proof I that the triangle PQR is equilateral. Thus we have $\angle ABC = \angle BCD = \cdots = \angle FAB = 120°$.

Second Day
9:00 – 13:30 July 14, 2003

4. Let $ABCD$ be an inscribed quadrilateral. Let P, Q and R be the feet of the perpendiculars from D to the lines BC, CA and AB respectively. Show that $PQ = QR$ if and only if the bisectors of $\angle ABC$ and $\angle ADC$ meet on AC.

Proof By Simson's Theorem, we know that P, Q, R are collinear. Moreover, since $\angle DPC$ and $\angle DQC$ are right angles, the points D, P, Q, C are concyclic and so $\angle DCA = \angle DPQ = \angle DPR$. Similarly, since D, Q, R, A are concyclic, we have $\angle DAC = DRP$. Therefore $\triangle DCA \backsim \triangle DPR$.

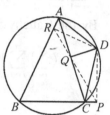

Likewise, $\triangle DAB \backsim \triangle DQP$ and $\triangle DBC \backsim \triangle DRQ$. Then

$$\frac{DA}{DC} = \frac{DR}{DP} = \frac{DB \cdot \dfrac{QR}{BC}}{DB \cdot \dfrac{PQ}{BA}} = \frac{QR}{PQ} \cdot \frac{BA}{BC}.$$

Thus $PQ = QR$ if and only if $\dfrac{DA}{DC} = \dfrac{BA}{BC}$.

Now the bisectors of the angles ABC and ADC divide AC in the ratios of $\dfrac{BA}{BC}$ and $\dfrac{DA}{DC}$ respectively. This completes the proof.

5. Let n be a positive integer and x_1, x_2, $\cdots$, x_n be real numbers with $x_1 \leqslant x_2 \leqslant \cdots \leqslant x_n$.

(a) Prove that

$$\left(\sum_{i=1}^{n} \sum_{j=1}^{n} |x_i - x_j|\right)^2 \leqslant \frac{2(n^2 - 1)}{3} \sum_{i=1}^{n} \sum_{j=1}^{n} (x_i - x_j)^2.$$

(b) Show that the equality holds if and only if x_1, x_2, $\cdots$, x_n is an arithmetic sequence.

Proof (a) Since both sides of the inequality are invariant when we subtract the same number from all $x_i's$, we may assume without loss of generality that $\sum\limits_{i=1}^{n} x_i = 0$.

We have

$$\sum_{i=1}^{n}\sum_{j=1}^{n}|x_i - x_j| = 2\sum_{i<j}(x_j - x_i) = 2\sum_{i=1}^{n}(2i - n - 1)x_i.$$

By the Cauchy-Schwarz inequality, we have

$$\left(\sum_{i=1}^{n}\sum_{j=1}^{n}|x_i - x_j|\right)^2 \leqslant 4\sum_{i=1}^{n}(2i - n - 1)^2\sum_{i=1}^{n}x_i^2$$

$$= \frac{4n(n+1)(n-1)}{3}\sum_{i=1}^{n}x_i^2.$$

On the other hand, we have

$$\sum_{i=1}^{n}\sum_{j=1}^{n}(x_i - x_j)^2 = n\sum_{i=1}^{n}x_i^2 - \sum_{i=1}^{n}x_i\sum_{j=1}^{n}x_j + n\sum_{j=1}^{n}x_j^2$$

$$= 2n\sum_{i=1}^{n}x_i^2.$$

Therefore

$$\left(\sum_{i=1}^{n}\sum_{j=1}^{n}|x_i - x_j|\right)^2 \leqslant \frac{2(n^2-1)}{3}\sum_{i=1}^{n}\sum_{j=1}^{n}(x_i - x_j)^2.$$

(b) If the equality holds, then there exists a real number k, such that $x_i = k(2i - n - 1)$, which means that $x_1, x_2, \cdots, x_n$ is an arithmetic sequence.

On the other hand, suppose that $x_1, x_2, \cdots, x_n$ is an arithmetic sequence with common difference d. Then we have

$$x_i = \frac{d}{2}(2i - n - 1) + \frac{x_1 + x_n}{2}.$$

Subtract $\dfrac{x_1 + x_n}{2}$ from every x_i, we obtain $x_i = \dfrac{d}{2}(2i - n - 1)$ and $\sum\limits_{i=1}^{n} x_i = 0$, from which the equality follows.

Let p be a prime number. Prove that there exists a prime number q such that for every integer n, the number $n^p - p$ is not divisible by q.

Proof Since $\dfrac{p^p - 1}{p - 1} = 1 + p + p^2 + \cdots + p^{p-1} \equiv p + 1 \pmod{p^2}$, we

can get a prime divisor q of $\dfrac{p^p - 1}{p - 1}$ such that $q \not\equiv 1 \pmod{p^2}$. This q

is what we wanted. The proof is given as follows.

Assume that there exists an integer n such that $n^p \equiv p \pmod{q}$.

Then we have $n^{p^2} \equiv p^p \equiv 1 \pmod{q}$ by the definition of q. On the other hand, from Fermat's little Theorem, $n^{q-1} \equiv 1 \pmod{q}$, because q is a prime. Since $p^2 \mid q - 1$, we have $\gcd(p^2, q-1) \mid p$, which leads to $n^p \equiv 1 \pmod{q}$. Hence we have $p \equiv 1 \pmod{q}$. However, this implies $1 + p + \cdots + p^{p-1} \equiv p \pmod{q}$. From the definition of q, we have $0 \equiv \dfrac{p^p - 1}{p - 1} = 1 + p + \cdots + p^{p-1} \equiv p \pmod{q}$, but this leads to a contradiction.

2004 (Athens, Greece)

The 45th IMO (International Mathematical Olympiad) was hosted by Greece in Athens during July 4 - 18 in 2004. 84 countries and 486 contestants participated. 45 gold medals, 78 silver medals, and 120 bronze medals were awarded.

The leader of Chinese IMO2004 team was Chen Yonggao, deputy leader was Xiong Bin, observers were Wu Jianping, Wang Shuguo. In IMO2004, China came first among the nations with six golds. Here are the results:

Huang Zhiyi	the High School Attached to South China Normal University	41 points	gold medal
Zhu Qingsan	the High School Attached to South China Normal University	38 points	gold medal
Li Xianyin	the High School Attached to Hunan Normal University	37 points	gold medal
Lin Yuncheng	Shanghai High School	35 points	gold medal
Peng Minyu	No. 1 High School of Yingtan, Jiangxi	35 points	gold medal
Yang Shiwu	Hubei Huanggang High School	34 points	gold medal

First Day
9:00 – 13:30 July 12, 2004

Let ABC be an acute-angled triangle with $AB \neq AC$. The circle with diameter BC intersects the sides AB and AC at M and N respectively. Denote by O the midpoint of the side BC. The bisectors of the angles BAC and MON intersect at R. Prove that the circumcircles of the triangles BMR and CNR have a common point lying on the side BC.

Solution We first show that the points A, M, R, N are concyclic. Since ABC is an acute-angled triangle, M and N are on the line segments AB and AC respectively. Let R_1 be the point such that the points A, M, R_1, N are concyclic, where R_1 is on the ray AR. Since AR_1 bisects $\angle BAC$, we have $R_1M = R_1N$. Since M and N lie on the circle with centre O, we have $OM = ON$. It follows from $OM = ON$ and $R_1M = R_1N$ that R_1 is on the bisector of $\angle MON$. Since $AB \neq AC$, the bisectors of the angles BAC and MON intersect at the unique point R, and so $R_1 = R$, or A, M, R, N are concyclic.

Let the bisector of $\angle BAC$ meet BC at K. Since the points B, C,

N, M are concyclic, $\angle MBC = \angle ANM$. Moreover, since A, M, R, N are concyclic, $\angle ANM = \angle MRA$. This implies $\angle MBK = \angle MRA$. Therefore, the points B, M, R, K are concyclic. Using the same argument as above, we obtain that C, N, R, K are concyclic. This completes the solution.

Find all polynomials $P(x)$ with real coefficients, which satisfy the equation

$$P(a-b) + P(b-c) + P(c-a) = 2P(a+b+c)$$

for all real numbers a, b, c such that $ab + bc + ca = 0$.

Solution Let $P(x)$ satisfy the given equation.

If $a = b = c = 0$, then $P(0) = 0$.

If $b = c = 0$, then $P(-a) = P(a)$ for all real a.

Hence $P(x)$ is even. Without loss of generality, we may assume that

$$P(x) = a_n x^{2n} + \cdots + a_1 x^2, \ a_n \neq 0.$$

If $b = 2a$, $c = -\dfrac{2}{3}a$, we have that

$$P(-a) + P\left(\frac{8}{3}a\right) + P\left(-\frac{5}{3}a\right) = 2P\left(\frac{7}{3}a\right),$$

or $a_n\left[1 + \left(\dfrac{8}{3}\right)^{2n} + \left(\dfrac{5}{3}\right)^{2n} - 2\left(\dfrac{7}{3}\right)^{2n}\right]a^{2n} + \cdots + a_1\left[1 + \left(\dfrac{8}{3}\right)^{2} + \left(\dfrac{5}{3}\right)^{2} - 2\left(\dfrac{7}{3}\right)^{2}\right]a^2 = 0$ for all $a \in R$. Then all coefficients of the polynomial with variable a are 0.

If $n \geqslant 3$, it follows from $8^6 = 262\ 144 > 235\ 298 = 2 \times 7^6$ that $\left(\dfrac{8}{7}\right)^{2n} \geqslant \left(\dfrac{8}{7}\right)^{6} > 2$. This implies

$$1 + \left(\frac{8}{3}\right)^{2n} + \left(\frac{5}{3}\right)^{2n} - 2\left(\frac{7}{3}\right)^{2n} > 0.$$

Hence $n \leqslant 2$. Let $P(x) = \alpha x^4 + \beta x^2$, with α, $\beta \in R$. We now show that $P(x) = \alpha x^4 + \beta x^2$ satisfies the given equation. Let a, b, c be real numbers satisfying $ab + bc + ca = 0$. Then

$$(a-b)^4 + (b-c)^4 + (c-a)^4 - 2(a+b+c)^4$$

$$= \sum (a^4 - 4a^3 b + 6a^2 b^2 - 4ab^3 + b^4) - 2(a^2 + b^2 + c^2)^2$$

$$= \sum (a^4 - 4a^3 b + 6a^2 b^2 - 4ab^3 + b^4) -$$

$$2a^4 - 2b^4 - 2c^4 - 4a^2 b^2 - 4a^2 c^2 - 4b^2 c^2$$

$$= \sum (-4a^3 b + 2a^2 b^2 - 4ab^3)$$

$$= -4a^2 (ab + ca) - 4b^2 (bc + ab) - 4c^2 (ca + bc) +$$

$$2(a^2 b^2 + b^2 c^2 + c^2 a^2)$$

$$= 4a^2 bc + 4b^2 ca + 4c^2 ab + 2a^2 b^2 + 2b^2 c^2 + 2c^2 a^2$$

$$= 2(ab + bc + ca)^2 = 0,$$

$$(a-b)^2 + (b-c)^2 + (c-a)^2 - 2(a+b+c)^2$$

$$= \sum (a^2 - 2ab + b^2) - 2 \sum a^2 - 4 \sum ab$$

$$= 0.$$

Hence $P(x) = \alpha x^4 + \beta x^2$ satisfies the given equation.

Define a *hook* to be a figure made up of six unit squares as shown in the diagram or any of the figures obtained by applying rotations and reflections to this figure.

　　Determine all $m \times n$ rectangles that can be covered with hooks so that

- the rectangle is covered without gaps and without overlaps;
- no part of a hook covers area outside the rectangle.

Solution I　　m and n should be the positive integers and should satisfy one of the following conditions: (1) $3 \mid m$ and $4 \mid n$ (or vice versa); (2) one of m and n is divisible by 12 and one is not less than 7.

A figure is obtained by applying rotations and reflections to another figure. We regard the two figures as equivalent.

Label the six unit squares of the hook as shown below. The shaded square must belong to another hook, and it is adjacent to only one square of this other hook. Then the only possibility of the shaded square is 1 or 6.

(i) If it is 6, two hooks form a 3×4 rectangle. We call it ①.

(ii) If it is 1, there are two cases.

It is easy to see that the shaded square cannot be covered in the first diagram as shown below. Hence the latter is true. We call it ②.

Thus, in a tessellation, all hooks are matched into pairs. Each pair forms ① or ②.

There are 12 squares in ① and ②. Hence $12 \mid mn$.

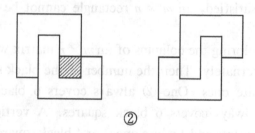

②

Now we consider three cases, separately.

(1) $3 \mid m$ and $4 \mid n$ (or vice versa)

Without loss of generality, we may assume $m = 3m_0$ and $n = 4n_0$. Then $m_0 n_0$ rectangles of the type ① form an $m_0 \times n_0$ rectangle. Since two hooks cover a 3×4 rectangle, an $m \times n$ rectangle can be covered with hooks.

(2) $12 \mid m$ or $12 \mid n$. Without loss of generality, we may assume $12 \mid m$.

If $3 \mid n$ or $4 \mid n$, the question reduces to (1).

Assume that n is not divisible by 3 nor by 4. If a tessellation exists, then there is at least one of ① and ② in it, so $n \geqslant 3$. Hence $n \geqslant 5$ because $3 \nmid n$ and $4 \nmid n$. Since the square at the corners can belong to either ① or ②, it follows from $n \geqslant 5$ that the squares at the adjacent corners cannot belong to the same type ① or ②. Hence $n \geqslant 6$. Since n is not divisible by 3 and 4, $n \geqslant 7$.

We now show that if $n \geqslant 7$ and n is not divisible by 3 and 4, a tessellation exists.

If $n \equiv 1 \pmod 3$, then $n = 4 + 3t$ ($t \in N^*$). Together with (1), we have that if $12 \mid m$, an $m \times 3t$, rectangle and an $m \times 4$ rectangle can be covered with hooks. So the problem can be solved.

If $n \equiv 2 \pmod 3$, $n = 8 + 3t$ ($t \in N^*$). Together with (1), we have that if $12 \mid m$, each of $m \times 8$ and $m \times 3t$ rectangles can be covered with hooks. The problem is solved.

(3) $12 \mid mn$, but neither m nor n is divisible by 4. Now $2 \mid m$, $2 \mid n$. We may assume without loss of generality that $m = 6m_0$, $n = 2n_0$, neither m_0 nor n_0 is divisible by 2. We will prove that if these conditions are satisfied, an $m \times n$ rectangle cannot be covered with hooks.

Consider coloring the columns of an $m \times n$ matrix with black and white colors alternately. Then the number of the black squares equals that of the white ones. One ② always covers 6 black squares. A horizontal ① always covers 6 black squares. A vertical ① covers either 8 black squares and 4 white ones, or 4 black squares and 8 white ones. Since the number of the black squares equals that of the white ones, the number of ① is the same in the preceding two cases. Hence the total number of a vertical ① is even. Using the same argument as above (coloring the rows alternately), we obtain that the total number of a horizontal ① is even.

Consider classifying the squares of the $m \times n$ rectangle into 4 types marked 1, 2, 3, and 4 as shown below. The number of squares of each type is equal to $\frac{mn}{4}$.

1	2	3	4	1	2	...	1	2
3	4	1	2	3	4	...	3	4
1	2	3	4	1	2	...	1	2
						...		
						...		
:	:	:	:	:	:		:	:
3	4	1	2	3	4		3	4

From the two diagrams,

a	b	c	d
c	d	a	b
a	b	c	d

a	b	c
c	d	a
a	b	c
c	d	a

we obtain that the number of a and c covered by ① is the same, so is for b and d. Hence the number of squares of type 1 covered by ① equals that of type 3.

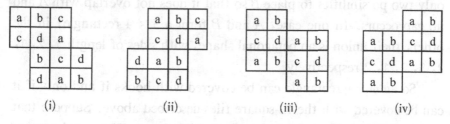

(i) (ii) (iii) (iv)

The number of squares of type 1 covered by (i) or (ii) equals that of type 3. The difference between the number of squares of type 1 and type 3 covered by (iii) or (iv) is 2. There are two cases: the number of squares of type 1 is 2 more than that of type 3, or vice versa. Since the number of squares of type 1 equals that of type 3 in the rectangle, the frey nancy of the two cases is the same. Hence the total number of (iii) and (iv) is even.

Consider classifying the squares of the $m \times n$ rectangle as shown below.

1	3	1		
2	4	2		
3	1	3		
4	2	4		

Similarly, the total number of (i) and (ii) is even. Then the number of ② is even. So there is an ever number of ① and ②. Hence $24 \mid m \times n$, contrary to the assumption that neither m nor n is divisible by 4.

Solution II m and n should be the positive integers and should satisfy one of the following conditions: (1) $3 \mid m$ and $4 \mid n$ (or vice versa); (2) $12 \mid m$, $n \neq 1, 2, 5$ (or vice versa).

Consider a covering of an $m \times n$ rectangle satisfying the conditions. For any hook A, there is a unique hook B covering the "inner" square of A with one of its "tailend" squares. In turn, the "inner" square of B must be covered by an "tailend" square of A. Thus, in a tessellation, all hooks are matched into pairs. There are only two possibilities to place B so that it does not overlap with A and no gap occurs. In one case, A and B form a 3×4 rectangle; in the other, their union is an octagonal shape, with sides of length 3, 2, 1, 2, 3, 2, 1, 2 respectively.

So an $m \times n$ rectangle can be covered with hooks if and only if it can be covered with the 12-square tiles described above. Suppose that such a tessellation exists; then mn is divisible by 12. We now show that one of m and n is divisible by 4.

Assume on the contrary that this is not the case. Then m and n are both even, because mn is divisible by 4. Imagine that the rectangle is divided into unit squares, with the rows and columns labeled $1, \cdots$, m and $1, \cdots, n$. Write 1 in the square (i, j) if exactly one of i and j is divisible by 4, and 2, if i and j are both divisible by 4. Since the number of squares in each row and column is even, the sum of all numbers written is also even. Now, it is easy to check that a 3×4 rectangle always covers numbers with sum 3 or 7; and the other 12-square shape always covers numbers with sum 5 or 7. Consequently, the total number of 12-square shapes is even. But then mn is divisible by 24, and hence by 8, contrary to the assumption that m and n are not divisible by 4.

Notice also that neither m nor n can be 1, 2 or 5 (any attempt to place tiles along a side of length 1, 2 or 5 fails). We infer that if a tessellation is possible, then one of m and n is divisible by 3, one is divisible by 4, and m, $n \notin \{1, 2, 5\}$.

Conversely, we shall prove that if these conditions are satisfied, then a tessellation is possible (using only 3×4 rectangles). The result is immediate if 3 divides m and 4 divides n (or vice versa). Let m be divisible by 12 and $n \notin \{1, 2, 5\}$ (or vice versa). Without loss of generality, we may assume that neither 3 nor 4 divides n. Then $n \geq 7$. In addition, between $n-4$ and $n-8$, at least one can be divisible by 3. Hence the rectangle can be partitioned into $m \times 3$ and $m \times 4$ rectangles, which are easy to cover, in fact with only 3×4 tiles again.

Second Day
9:00 – 13:00 July 13, 2004

⬤ Let $n \geq 3$ be an integer. Let t_1, t_2, $\cdots$, t_n be positive real numbers such that

$$n^2 + 1 > (t_1 + t_2 + \cdots + t_n)\left(\frac{1}{t_1} + \frac{1}{t_2} + \cdots + \frac{1}{t_n}\right).$$

Show that t_i, t_j, t_k are the lengths of the sides of a triangle for all i, j, k with $1 \leq i < j < k \leq n$.

Solution Assume on the contrary that there exist three numbers among t_1, t_2, $\cdots$, t_n that do not form the sides of a triangle. Without loss of generality, we may assume that these three numbers are t_1, t_2,

t_3, and $t_1 + t_2 \leqslant t_3$. One has

$$(t_1 + \cdots + t_n)\left(\frac{1}{t_1} + \cdots + \frac{1}{t_n}\right)$$

$$= \sum_{1 \leqslant i < j \leqslant n}\left(\frac{t_i}{t_j} + \frac{t_j}{t_i}\right) + n$$

$$= \frac{t_1}{t_3} + \frac{t_3}{t_1} + \frac{t_2}{t_3} + \frac{t_3}{t_2} + \sum_{\substack{1 \leqslant i < j \leqslant n \\ (i,\,j) \notin \{(1,\,3),\,(2,\,3)\}}}\left(\frac{t_i}{t_j} + \frac{t_j}{t_i}\right) + n$$

$$\geqslant \frac{t_1 + t_2}{t_3} + t_3\left(\frac{1}{t_1} + \frac{1}{t_2}\right) + \sum_{\substack{1 \leqslant i < j \leqslant n \\ (i,\,j) \notin \{(1,\,3),\,(2,\,3)\}}} 2 + n$$

$$\geqslant \frac{t_1 + t_2}{t_3} + \frac{4t_3}{t_1 + t_2} + 2(C_n^2 - 2) + n$$

$$= 4\frac{t_3}{t_1 + t_2} + \frac{t_1 + t_2}{t_3} + n^2 - 4. \tag{1}$$

If $x = \dfrac{t_3}{t_1 + t_2}$, then $x \geqslant 1$, and $4x + \dfrac{1}{x} - 5 = \dfrac{(x-1)(4x-1)}{x} \geqslant 0$.

Together with (1), we obtain that

$$(t_1 + \cdots + t_n)\left(\frac{1}{t_1} + \cdots + \frac{1}{t_n}\right) \geqslant 5 + n^2 - 4 = n^2 + 1,$$

a contradiction. This completes the proof.

Remark   By the AM-GM inequality,

$$\left(\frac{1}{t_1} + \frac{1}{t_2}\right)(t_1 + t_2) \geqslant 2\sqrt{\frac{1}{t_1 t_2}} \cdot 2\sqrt{t_1 t_2} = 4,$$

which is the same as $\dfrac{1}{t_1} + \dfrac{1}{t_2} \geqslant \dfrac{4}{t_1 + t_2}$.

⑤ In a convex quadrilateral $ABCD$, the diagnal BD bisects neither the angle ABC nor the angle CDA. The point P lies inside $ABCD$ and satisfies

$$\angle PBC = \angle DBA \text{ and } \angle PDC = \angle BDA.$$

Prove that $ABCD$ is a cyclic quadrilateral if and only if $AP = CP$.

Solution I (i) Necessity.

Assume that $ABCD$ is a cyclic quadrilateral. Let the circle Γ be the circumcircle of quadrilateral $ABCD$. Extend BP and DP beyond P to meet the circle Γ at X and Y respectively.

Since DB does not bisect $\angle ABC$ and P lies inside $ABCD$, it follows from $\angle PBC = \angle DBA$ that $\overset{\frown}{CX} = \overset{\frown}{AD}$, $D \neq X$, and the points D and X lie in the same half-plane bounded by the line AC. Hence $DX /\!/ AC$. Similarly, $B \neq Y$ and $BY /\!/ AC$.

The points D, X, A, C, B, Y lie on the circle Γ, as mentioned above. So the points D and X, the points A and C, the points B and Y are symmetrical with respect to the perpendicular bisector l of AC respectively.

Since $P = DY \cap BX$, then P is on the line l, or $AP = CP$. This completes the proof of the necessity.

(ii) Sufficiency.

Lemma. *Let l be a fixed line and A, B, C be fixed points such that A and B, C lie in the different half-planes bounded by the line l. Assume that the point X lies on the line l. Let $\alpha(X)$ be the smallest angle of rotation from the line XA to the line l anticlockwise. Let $\beta(X)$ be the smallest angle of rotation from the line XC to the line XB anticlockwise. If $\alpha(X) = \beta(X)$, then X is called a good point. Prove that there are at most two good points.*

Proof of the lemma

We set up a coordinate system with the line l as the x-axis and the perpendicular of l as the y-axis.

Let $A(a, b)$, $B(c, d)$, $C(e, f)$, $X(x, 0)$. Then $A = a + bi$, $B = c + di$, $C = e + fi$, and $X = x$. Hence

$$A - X = a - x + bi,$$

$$(a - x + bi)(\cos\alpha(X) + i\sin\alpha(X))$$

$$= (a - x)\cos\alpha(X) - b\sin\alpha(X) + [(a - x)\sin\alpha(X) + b\cos\alpha(X)]i.$$

Rotate the line XA by $\alpha(X)$ anticlockwise, coinciding with l. So

$$(a-x)\cos\alpha(X)+b\cos\alpha(X)=0 \qquad (1)$$

Moreover, since $\overrightarrow{XC}=e-x+f\mathrm{i}$, $\overrightarrow{XB}=c-x+d\mathrm{i}$, then $\overrightarrow{XC}$ after being rotated by $\beta(X)$ anticlockwise becomes

$$(e-x+f\mathrm{i})(\cos\beta(X)+\mathrm{i}\sin\beta(X))$$

$$=(e-x)\cos\beta(X)-f\sin\beta(X)+[(e-x)\sin\beta(X)+f\cos\beta(X)]\mathrm{i}$$

which is parallel to $\overrightarrow{XB}$. Thus,

$$(c-x)[(e-x)\sin\beta(X)+f\cos\beta(X)]-d[(e-x)\cos\beta(X)-f\sin\beta(X)]=0,$$

and

$$[(c-x)(e-x)+df]\sin\beta(X)$$
$$+[(c-x)f-d(e-x)]\cos\beta(X)=0. \qquad (2)$$

Since $\sin\theta$ and $\cos\theta$ are not 0 at the same time, so

X is a good point $\Leftrightarrow \alpha(X)=\beta(X) \overset{(1)(2)}{\Leftrightarrow}$ the simultaneous equations in u and v.

$$\begin{cases}(a-x)u+bv=0,\\ [(c-x)(e-x)+df]u+[(c-x)f-d(e-x)]v=0\end{cases}$$

have the non-zevo solutions

$$\Leftrightarrow b[(c-x)(e-x)+df]+(a-x)[(c-x)f-d(e-x)]=0$$

$$\Leftrightarrow (b+d-f)x^2+(af+cf-bc-be-ad-ed)x$$

$$+bce+bdf+ade-acf=0.$$

Let $g(x)=(b+d-f)x^2+(af+cf-bc-be-ad-ed)x+bce+bdf+ade-acf$. If $b+d-f=0$, it follows from $b\neq 0$ that $d\neq f$. Hence BC and l are not parallel. Let BC intersect l at $T(t,0)$. We have that $\beta(T)=0$ and $\alpha(T)>0$. Hence T is not a good point. This implies $g(t)\neq 0$, so the equation $g(x)=0$ has at most two solutions.

Hence there are at most two good points.

This completes the proof of the lemma.

Let the points A, B, C, D be arranged clockwise and $AP = CP$. Let D^* be the point such that A, D^*, C, B are concyclic, where D^* lies on the ray BD. Since $AP = CP$ and BD bisects neither $\angle ABC$ nor $\angle ADC$, BP intersects the perpendicular bisector of AC at the unique point P.

Let $D^* R$ be the ray satisfying $\angle CD^* R = \angle AD^* B$ and $P^* = D^* R \cap BP$.

Together with (i), we have that $P^* A = P^* C$, or P^* is the intersection point of l and the perpendicular bisector of AC, so $P^* = P$. It follows from $\angle CD^* R = \angle AD^* B$ that $\angle AD^* B = \angle CD^* P$.

Replace l, A, B, C by BD, A, C, P. Then B, D, D^* are good points. Since $B \neq D$ and $B \neq D^*$, $D = D^*$. Hence A, B, C, D is concyclic.

This completes the proof of the sufficiency.

Combining (i) and (ii), we obtain that the conclusion is true.

Solution II We may assume without loss of generality that P lies in the rectangles ABC and BCD.

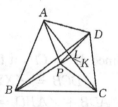

Assume that the quadrilateral $ABCD$ is cyclic. Let the lines BP and DP meet AC at K and L respectively. It follows from the given equalities and $\angle ACB = \angle ADB$, $\angle ABD = \angle ACD$ that the triangles DAB, DLC and CKB are similar. This implies $\angle DLC = \angle CKB$, so $\angle PLK = \angle PKL$. Hence $PK = PL$.

It follows from $\angle BDA = \angle PDC$ that $\angle ADL = \angle BDC$. Since $\angle DAL = \angle DBC$, the triangles ADL and BDC are similar. Hence

$$\frac{AL}{BC} = \frac{AD}{BD} = \frac{KC}{BC},$$

yielding $AL = KC$. It follows from $\angle DLC = \angle CKB$ that $\angle ALP = \angle CKP$. Moreover, since $PK = PL$ and $AL = KC$, the triangles ALP

and CKP are congruent. Hence $AP = CP$.

Conversely, assume that $AP = CP$. Let the circumcircle of the triangle BCP meet the lines CD and DP again at X and Y respectively.

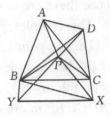

It follows from $\angle ADB = \angle PDX$ and $\angle ABD = \angle PBC = \angle PXC$ that the triangles ADB and PDX are similar. Hence

$$\frac{AD}{PD} = \frac{BD}{XD}$$

Since $\angle ADP = \angle ADB + \angle BDP = \angle PDX + \angle BDP = \angle BDX$, the triangles ADP and BDX are similar. Therefore,

$$\frac{BX}{AP} = \frac{BD}{AD} = \frac{XD}{PD}. \tag{1}$$

Since the points P, C, X, Y are concyclic, $\angle DPC = \angle DXY$ and $\angle DCP = \angle DYX$. Hence the triangles DPC and DXY are similar. Thus,

$$\frac{YX}{CP} = \frac{XD}{PD}. \tag{2}$$

Since $AP = CP$, it follows from (1) and (2) that $BX = YX$. Hence

$\angle DCB = \angle XYB = \angle XBY = \angle XPY = \angle PDX + \angle PXD = \angle ADB + \angle ABD = 180° - \angle BAD$. The above equality means that $ABCD$ is a cyclic quadrilateral.

⑩ We call a positive integer *alternating* if every two consecutive digits in its decimal representation are of different parity.

Find all positive integers n such that n has a multiple which is alternating.

Solution Ⅰ　　Lemma 1　　*If k is a positive integer, then there exist $0 \leqslant a_1, a_2, \cdots, a_{2k} \leqslant 9$ such that $a_1, a_3, \cdots, a_{2k-1}$ are odd integers, $a_2, a_4, \cdots, a_{2k}$ are even integers, and*

$$2^{2k+1} \mid \overline{a_1 a_2 \cdots a_{2k}} \quad (\textit{The decimal representation})$$

Proof of Lemma 1 by mathemations induction.

If $k = 1$, it follows from $8 \mid 16$ that the proposition is true.

Assume that if $k = n - 1$, the proposition is true.

When $k = n$, let $\overline{a_1 a_2 \cdots a_{2n-2}} = 2^{2n-1} t$ by the inductive hypothesis. The problem reduces to proving that there exist $1 \leqslant a, b \leqslant 9$ with a odd and b even such that $2^{2n+1} \mid \overline{ab} \times 10^{2n-2} + 2^{2n-1} t$, or $8 \mid \overline{ab} \times 5^{2n-2} + 2t$, or $8 \mid \overline{ab} + 2t$ in view of $5^{2n-2} \equiv 1 \pmod 8$.

It follows from $8 \mid 12 + 4$, $8 \mid 14 + 2$, $8 \mid 16 + 0$ and $8 \mid 50 + 6$ that the Lemma 1 is true.

Lemma 2   *If k is a positive integer, then there exists an alternative number $\overline{a_1 a_2 \cdots a_{2k}}$ with an even number $2k$ of digits such that a_{2k} is odd and $5^{2k} \mid \overline{a_1 a_2 \cdots a_{2k}}$, where a_1 can be 0, but $a_2 \neq 0$.*

Proof of Lemma 2 by mathematical induction.

If $k = 1$, it follows from $25 \mid 25$ that the proposition is true.

Assume that if $k = n - 1$, the proposition is true, or there exists an alternative multiple of $\overline{a_1 a_2 \cdots a_{2n-2}}$ satisfying $5^{2n-2} \mid \overline{a_1 a_2 \cdots a_{2n-2}}$.

When $k = n$, let $\overline{a_1 a_2 \cdots a_{2n-2}} = t \cdot 5^{2n-2}$. The problem reduces to proving that there exist $0 \leqslant a, b \leqslant 9$ with a even and b odd such that $5^{2n} \mid \overline{ab} \times 10^{2n-2} + t \cdot 5^{2n-2}$, or $25 \mid \overline{ab} \times 2^{2n-2} + t$.

Since 2^{2n-2} is coprime to 25, there exist $0 < \overline{ab} \leqslant 25$ such that $25 \mid \overline{ab} \times 2^{2n-2} + t$. If b is odd, between $\overline{ab}$ and $\overline{ab} + 50$, at least one satisfies that the highest-valued digit is even. If b is even, between $\overline{ab} + 25$ and $\overline{ab} + 75$, at least one satisfies that the highest-valued digit is even.

This completes the proof of Lemma 2.

Let $n = 2^\alpha 5^\beta t$, where t is coprime to 10 and $\alpha, \beta \in \mathbf{N}$. Assume that $\alpha \geqslant 2$ and $\beta \geqslant 1$. Let l be an arbitrary multiple of n. The last decimal digit is 0, and the digit in tens is even. Hence these n do not satisfy the required condition.

① When $\alpha = \beta = 0$, consider $21, 2\,121, 212\,121, \cdots$, $\underbrace{2\,121 \cdots 21}_{\text{number } k \text{ of } 21}, \cdots$. There must exist two of them congruent modulo n.

Without loss of generality, we may assume that $t_1 > t_2$, and

$$\underbrace{2\;121\cdots21}_{\text{number } t_1 \text{ of } 21} \equiv \underbrace{2\;121\cdots21}_{\text{number } t_2 \text{ of } 21} \pmod{n}.$$

Then

$$\underbrace{2\;121\cdots21}_{\text{number } t_1-t_2 \text{ of } 21}\underbrace{00\cdots0}_{\text{number } 2t_2 \text{ of } 0} \equiv 0 \pmod{n}$$

Hence

$$\underbrace{2\;121\cdots21}_{\text{number } t_1-t_2 \text{ of } 21} \equiv 0 \pmod{n},$$

because n is coprime to 10.

Now these positive integers n satisfy the required condition.

② When $\beta = 0$ and $\alpha \geqslant 1$, it follows from Lemma 1 that there exists an alternative number $\overline{a_1 a_2 \cdots a_{2k}}$ satisfying that $2^\alpha \mid \overline{a_1 a_2 \cdots a_{2k}}$. Consider

$$\overline{a_1 a_2 \cdots a_{2k}}, \; \overline{a_1 a_2 \cdots a_{2k} a_1 a_2 \cdots a_{2k}}, \; \cdots,$$

$$\underbrace{\overline{a_1 a_2 \cdots a_{2k} a_1 a_2 \cdots a_{2k} \cdots a_1 a_2 \cdots a_{2k}}}_{\text{number } l \text{ of sections}}, \; \cdots$$

There must exist two of them congruent modulo t. Without loss of generality, we may assume that $t_1 > t_2$, and

$$\underbrace{\overline{a_1 a_2 \cdots a_{2k} \cdots a_1 a_2 \cdots a_{2k}}}_{\text{number } t_1 \text{ of sections}} \equiv \underbrace{\overline{a_1 a_2 \cdots a_{2k} \cdots a_1 a_2 \cdots a_{2k}}}_{\text{number } t_2 \text{ of sections}} \pmod{t}$$

Since t is coprime to 10, $\overline{a_1 a_2 \cdots a_{2k} \cdots a_1 a_2 \cdots a_{2k}} \equiv 0 \pmod{t}$. Moreover, since t is coprime to 2,

$$2^\alpha t \mid \underbrace{\overline{a_1 a_2 \cdots a_{2k} \cdots a_1 a_2 \cdots a_{2k}}}_{\text{number } t_1-t_2 \text{ of sections}},$$

which is alternating.

③ When $\alpha = 0$, $\beta \geqslant 1$, it follows from lemma 2 that there exists an alternative multiple $\overline{a_1 a_2 \cdots a_{2k}}$ satisfying $5^\beta \mid \overline{a_1 a_2 \cdots a_{2k}}$ with a_{2k} odd. Using the same argument as in ②, we obtain that there exist $t_1 > t_2$ satisfying $t \mid \underbrace{\overline{a_1 a_2 \cdots a_{2k} \cdots a_1 a_2 \cdots a_{2k}}}_{\text{number } t_1-t_2 \text{ of sections}}$. Since t is coprime to 5,

$5^{\beta}t \mid \underbrace{\overline{a_1 a_2 \cdots a_{2k} \cdots a_1 a_2 \cdots a_{2k}}}_{\text{number } t_1 - t_2 \text{ of sections}}$. In addition, $\underbrace{\overline{a_1 a_2 \cdots a_{2k} \cdots a_1 a_2 \cdots a_{2k}}}_{\text{number } t_1 - t_2 \text{ of sections}}$ is

alternating, and the last decimal digit a_{2k} is odd.

④ When $\alpha = 1$ and $\beta \geqslant 1$, it follows from ③ that there exists an alternative number $\overline{a_1 a_2 \cdots a_{2k} \cdots a_1 a_2 \cdots a_{2k}}$ satisfying that a_{2k} is odd and $5^{\beta}t \mid \overline{a_1 a_2 \cdots a_{2k} \cdots a_1 a_2 \cdots a_{2k}}$. Hence $2 \cdot 5^{\beta}t \mid \overline{a_1 a_2 \cdots a_{2k} \cdots a_1 a_2 \cdots a_{2k} 0}$ which is alternating.

In conclusion, if n is not divisible by 20, then these positive integers n satisfy the required condition.

Solution Ⅱ n should be the positive integers and should satisfy that n is not divisible by 20.

(1) Assume that $20 \mid n$. Select a multiple of n arbitrarily. Denote $(a_k a_{k-1} \cdots a_1)_{10}$, where $a_k \neq 0$. We have $20 \mid (a_k a_{k-1} \cdots a_1)_{10}$. Hence $a_1 = 0$, $2 \mid (a_k a_{k-1} \cdots a_2)_{10}$. This implies that a_2 is even. Therefore, $(a_k a_{k-1} \cdots a_1)_{10}$ is not alternating.

(2) We will prove that if n is a positive integer and is not divisible by 20, then n must have a multiple which is alternating.

We will show three lemmas as follows. Then we divide the proof into four cases.

Lemma 1   *If the positive integer n is coprime to* 10, *then for any* $l \in \mathbf{N}$, *there exists* $k \in \mathbf{N}$ *such that*

$$\underbrace{1 \underbrace{00 \cdots 0}_{\text{number } l \text{ of } 0} 1 \underbrace{00 \cdots 0}_{\text{number } l \text{ of } 0} 1 \cdots 1 \underbrace{00 \cdots 0}_{\text{number } l \text{ of } 0} 1}_{\text{total number } k \text{ of } 1}$$

is a multiple of n.

Proof of Lemma 1   Consider the numbers $x_1 = 1$, $x_2 = 1 \underbrace{00 \cdots 0}_{\text{number } l \text{ of } 0} 1$, $\cdots$,

$$x_m = 1 \underbrace{00 \cdots 0}_{\text{number } l \text{ of } 0} 1 \cdots 1 \underbrace{00 \cdots 0}_{\text{number } l \text{ of } 0} 1, \cdots.$$
$$\underbrace{}_{\text{number } m \text{ of } 1}$$

There must exist two of them congruent modulo n. Without loss of generality, we may assume that

$$x_s \equiv x_t (\mathrm{mod}\ n),\ s > t \geqslant 1.$$

Then $n \mid x_s - x_t$. Moreover, since $x_s - x_t = x_{s-t} \cdot 10^{t(l+1)}$ and n is coprime to 10, $n \mid x_{s-t}$, $s - t > 0$ and x_{s-t} is a positive integer. This completes the proof of Lemma 1.

Lemma 2 *For any* $m \in \mathbf{N}$, *there always exists an alternative number* s_m *with* m *digits such that its first digit can be* 0, *and its last decimal digit is* 5, *and* $5^m \mid s_m$.

Proof of Lemma 2 by inductive construction. Start with $s_1 = 5$. Suppose that $s_m = (a_m a_{m-1} \cdots a_1)_{10}$ is alternating, where $a_1 = 5$ and a_m can be 0. In addition, $5^m \mid s_m$. Let $s_m = 5^m$. Denote

$$A = \begin{cases} \{0,\ 2,\ 4,\ 6,\ 8\}, & \text{when } a_m \text{ is odd,} \\ \{1,\ 3,\ 5,\ 7,\ 9\}, & \text{when } a_m \text{ is even.} \end{cases}$$

Any two of them in A are not congruent modulo 5, and 2^m is coprime to 5. So any two of the numbers in $\{2^m x \mid x \in A\}$ are not congruent modulo 5. Select $x \in A$ such that $2^m x \equiv -l \pmod 5$. Then $5 \mid 2^m x + l$. Let $s_{m+1} = (x a_m a_{m-1} \cdots a_1)_{10}$. Then s_{m+1} is an alternative number with $m+1$ digits, where $a_1 = 5$, and its first digit can be 0. In addition, $s_{m+1} = x10^m + s_m = 5^m(2^m x + l)$ is a multiple of 5^{m+1}. This completes the proof of Lemma 2.

Lemma 3 *For any integer* $m \in \mathbf{N}$, *there always exists an alternative number* t_m *with* m *digits such that its last decimal digit is* 2, *and*

$$2^{2k+1} \parallel t_{2k+1},\ 2^{2k+3} \parallel t_{2k+2}.$$

Proof of Lemma 3 by inductive construction. Start with $t_1 = 2$. Suppose that $t_{2k+1} = (a_{2k+1} a_{2k} \cdots a_1)_{10}$ is alternating, where $a_1 = 2$, and $2^{2k+1} \parallel t_{2k+1}$. It follows from the property of the alternative numbers that $a_{2k+1} \equiv a_1 \equiv 0 \pmod 2$.

Denote $A = \{1,\ 3,\ 5,\ 7\}$. Any two of them in A are not congruent modulo 8, and 5^{2k+1} is coprime to 8. So any two numbers of $\{5^{2k+1} x \mid x \in A\}$ are not congruent modulo 8. Now diride the four numbers in $\{5^{2k+1} x \mid x \in A\}$ by 8 respectively. The original sentence means the 4 numbers are divisible by 8. Since $5^{2k+1}x$ with $x \in A$ is odd, the remainders are 1, 3, 5, 7.

Let $t_{2k+1} = 2^{2k+1}l$ with l odd. Select $x \in A$ such that $5^{2k+1}x \equiv -l+4 \pmod{8}$. Then $2^2 \parallel 5^{2k+1}x+l$. Let $t_{2k+2} = (xa_{2k+1}\cdots a_1)_{10}$. Then t_{2k+2} is alternating with $2k+2$ digits. In addition, $a_1 = 2$ and $t_{2k+2} = x10^{2k+1}+t_{2k+1} = 2^{2k+1}(5^{2k+1}x+l)$.

Since $2^2 \parallel 5^{2k+1}x+l$, we have $2^{2k+3} \parallel t_{2k+2}$.

Moreover, suppose that $t_{2k+2} = (a_{2k+2}\cdots a_1)_{10}$, where $a_1 = 2$, and $2^{2k+3} \parallel t_{2k+2}$. Let $t_{2k+3} = (4a_{2k+2}\cdots a_1)_{10}$. Then t_{2k+3} is an alternating number with $2k+3$ digits, and $t_{2k+3} = 5^{2k+2}2^{2k+4}+t_{2k+2}$.

Since $2^{2k+3} \parallel t_{2k+2}$, we have $2^{2k+3} \parallel t_{2k+3}$. This completes the proof of Lemma 3.

Next, we discuss the four cases.

(1) If n is coprime to 10, it follows from Lemma 1 that there exists $k \in N^*$ such that $\underbrace{10\ 101\cdots 101}_{\text{number } k \text{ of } 1}$ is a multiple of n. The conclusion is equivalent to selecting $l = 1$ in Lemma 1.

(2) If n is not divisible by 5, and n is divisible by 2, let $n = 2^m n_0$ such that n_0 is not divisible by 2, then n_0 is coprime to 10. Select $m_0 > m$ with m_0 even. It follows from Lemma 3 that there exists an alternative number $t_{m_0} = (a_{m_0}\cdots a_1)_{10}$ with m_0 digits, where $a_1 = 2$. In addition, $2^{m_0+1} \parallel t_{m_0}$. Hence $2^m \mid t_{m_0}$. It follows from Lemma 1 that there exists $k \in N^*$ such that

$$1\ \underbrace{00\cdots 0}_{\text{number } m_0-1 \text{ of } 0}\ 1\ \underbrace{00\cdots 0}_{\text{number } m_0-1 \text{ of } 0}\ \underbrace{1\cdots 1\ \underbrace{00\cdots 0}_{\text{number } m_0-1 \text{ of } 0}\ 1}_{\text{number } k \text{ of } 1}$$

is a multiple of n_0. Let

$$P = \underbrace{a_{m_0}a_{m_0-1}\cdots a_1 a_{m_0}\cdots a_1\cdots a_{m_0}\cdots a_1}_{\text{number } k \text{ of sections of } ``a_{m_0}\cdots a_1"}.$$

Then P is alternating, and

$$P = t_{m_0} \cdot 1 \underbrace{00 \cdots 0}_{\substack{number \ m_0-1 \ of \ 0}} 1 \cdots 100 \cdots 01$$

$$\underbrace{\qquad\qquad\qquad\qquad\qquad}_{number \ k \ of \ 1}$$

can be divisible by $2^m n_0$, or $n \mid P$.

(3) If n is divisible by 5, and n is not divisible by 2, let $n = 5^m n_0$ such that n_0 is not divisible by 5, then n_0 is coprime to 10. Select $m_0 >$ m with m_0 even. It follows from Lemma 2 that there exist an alternative number $s_{m_0} = (a_{m_0} \cdots a_1)_{10}$ with m_0 digits, where $a_1 = 5$ and a_{m_0} can be 0. In addition, $5^{m_0} \mid s_{m_0}$. Hence $5^m \mid s_{m_0}$. Using the same argument as (2), there exists an alternative number P such that $5^m n_0 \mid P$ and the last decimal digit $a_1 = 5$.

(4) If n is divisible by 10, and n is not divisible by 20, let $n = 10 n_0$ with n_0 odd, then n_0 satisfies the assumption in case(1) or in case (3). It follows from (1) and (3) that there exists $(a_k \cdots a_1)_{10}$ an alternative multiple of n_0, where $a_1 = 1$ or 5. Now Let $P = (a_k \cdots a_1 0)_{10}$. Then P is an alternative multiple of n.

In conclusion, n should be a positive integer and should not be divisible by 20.

2005   (Mérida, Mexico)

The 46th IMO (International Mathematical Olympiad) was hosted by Mérida in Mexico during July 8 - 18 in 2005. 92 countries and 513 contestants participated.

The leader of Chinese IMO2005 team was Xiong Bin who was from East China Normal University, deputy leader was Wang Jianwei who was from University of Science and Technology of China, observer was Chen Jinhui who was from the High School Attached to Fudan University. In IMO2005, China came first among the nations with total points of 235. Here are the results:

Ren Qingchun	Yaohua Middle School, Tianjin	42 points	gold medal
Diao Hansheng	the Second Middle School attached to East China Normal University	42 points	gold medal
Luo Ye	the High School Attached to Jiangxi Normal University	42 points	gold medal
Shao Xuancheng	High School Affiliated to Fudan University	42 points	gold medal
Kang Jiayin	Shenzhen Middle School	35 points	gold medal
Zhao Tongyuan	No. 2 Middle School of Shijiazhuang, Hebei	32 points	silver medal

First Day
9:00 – 13:30 July 13, 2005

Six points are chosen on the sides of an equilateral triangle ABC: A_1, A_2 on BC, B_1, B_2 on CA, and C_1, C_2 on AB. These points are the vertices of a convex hexagon $A_1 A_2 B_1 B_2 C_1 C_2$ with sides of equal length. Prove that the lines $A_1 B_2$, $B_1 C_2$ and $C_1 A_2$ are concurrent. (proposed by Romania, average score 2.61.)

Solution (posed by Diao Hansheng)

Assume $A_1 A_2 = d$, and $AB = a$. Construct an equilateral triangle $A_0 B_0 C_0$ with side of length $a - d$. Points A', B', C' are chosen on the sides of triangle $A_0 B_0 C_0$ such that $A'C_0 = A_2 C$, $B'A_0 = B_2 A$, and $C'B_0 = C_2 B$.

Therefore,

$$A'B_0 = a - d - A'C_0 = BC - A_1 A_2 - A_2 C = BA_1.$$

Similarly,

$$B'C_0 = B_1C, \; C'A_0 = C_1A.$$

Since

$$\angle B_1CA_2 = \angle B'C_0A', \angle B_2AC_1$$

$$= \angle B'A_0C', \angle C_2BA_1 = \angle C'B_0A',$$

thus

$$\triangle CB_1A_2 \cong \triangle C_0B'A', \; \triangle AB_2C_1 \cong \triangle A_0B'C',$$

$$\triangle C_2BA_1 \cong \triangle B_0C'A',$$

which implies that $B'C' = C'A' = A'B' = d$ and triangle $A'B'C'$ is equilateral.

So

$$\angle AB_2C_1 = \angle A_0B'C' = 180° - \angle C'B'A' - \angle A'B'C_0$$

$$= 120° - \angle A'B'C_0,$$

$$\angle C_1B_2B_1 = \angle B_1A_2A_1.$$

In view of $B_2C_1 = B_1B_2 = A_2B_1 = A_1A_2 = d$, triangles $C_1B_2B_1$ and $B_1B_2A_1$ are congruent, implying that $B_1C_1 = A_1B_1$. Together with $C_1C_2 = A_1C_2$, we show that C_2B_1 is the perpendicular bisector of A_1C_1 and C_2B_1 is the height of triangle $A_1B_1C_1$ on side A_1C_1. Similarly, C_1A_2 and A_1B_2 are the altitudes of triangle $A_1B_1C_1$ to the sides A_1B_1 and B_1C_1 respectively.

Therefore the lines A_1B_2, B_1C_2 and C_1A_2 are concurrent.

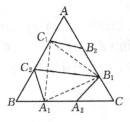 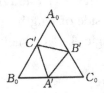

🔷 Let a_1, a_2, $\cdots$ be a sequence of integers with infinitely many positive terms and infinitely many negative terms. Suppose that for each positive integer n, the numbers a_1, a_2, $\cdots$, a_n leave n different remainders on division by n. Prove that each integer occurs exactly once in the sequence. (proposed by Holland, average score 3.05.)

Solution (posed by Ren Qingchun)

First we will show that every positive integer will occur at most once in the sequence. Note that if $a_i = a_j = k$ ($i < j$), then two numbers among a_1, a_2, $\cdots$, a_j, say a_i, a_j, are congruent modulo j, which is impossible.

Let x_k, y_k be the greatest and the smallest number among a_1, a_2, $\cdots$, a_k respectively. Then $x_k - y_k \leqslant k - 1$. For if $x_k - y_k \geqslant k$, without lose of generality, $a_i = x_k$, $a_j = y_k$, $a_i - a_j = l \geqslant k$, then i, $j \leqslant k \leqslant l$. Therefore two numbers among a_1, a_2, $\cdots$, a_j, say a_i, a_j, are congruent modulo l, which is impossible.

Now we will show that for every integer $t(y_k \leqslant t \leqslant x_k)$, there exists an integer $s(1 \leqslant s \leqslant k)$ such that $a_s = t$. For, if a_1, a_2, $\cdots$, $a_k \in \{u \in \mathbf{Z} \mid y_k \leqslant u \leqslant x_k, u \neq t\}$, then the sequence has $x_k - y_k$ different values at most. Note that $x_k - y_k \leqslant k - 1 < k$. Therefore two numbers among a_1, a_2, $\cdots$, a_k have the same value, which is a contradiction to the above argument.

Now for any integer m, since there are infinity many negative and positive numbers in the sequence, it is trivial to see that there exists a positive integer p such that $a_p > \mid m \mid$. By a similar argument, there exists a positive integer q such that $a_q < -\mid m \mid$. Denote $r = \max\{p, q\}$, then $x_r > \mid m \mid$, $y_r < -\mid m \mid$, i. e. $y_r < m < x_k$. The above arguments lead us to conclude that there exists a positive integer s such that $a_s = m$. We conclude that every number must appear exactly once in the sequence.

Let x, y and z be positive real numbers such that $xyz \geqslant 1$. Prove that

$$\frac{x^5 - x^2}{x^5 + y^2 + z^2} + \frac{y^5 - y^2}{y^5 + z^2 + x^2} + \frac{z^5 - z^2}{z^5 + x^2 + y^2} \geqslant 0. \quad \text{(proposed}$$

by Korea(R. O.))

Solution (posed by Kang Jiayin)

We shall prove

$$\sum \frac{x^5}{x^5 + y^2 + z^2} \geqslant 1 \geqslant \sum \frac{x^2}{x^5 + y^2 + z^2}. \qquad \text{①}$$

Suppose $xyz = d^3 \geqslant 1$. Let $x = x_1 d$, $y = y_1 d$, $z = z_1 d$, then $x_1 y_1 z_1 = 1$,

$$\sum \frac{x^5}{x^5 + y^2 + z^2} = \sum \frac{x_1^5 d^3}{x_1^5 d^3 + y_1^2 + z_1^2}$$

$$= \sum \frac{x_1^5}{x_1^5 + \frac{1}{d^3}(y_1^2 + z_1^2)}$$

$$\geqslant \sum \frac{x_1^5}{x_1^5 + y_1^2 + z_1^2},$$

$$\sum \frac{x^2}{x^5 + y^2 + z^2} = \sum \frac{x_1^2 d^2}{x_1^5 d^5 + y_1^2 d^2 + z_1^2 d^2}$$

$$= \sum \frac{x_1^2}{x_1^5 d^3 + y_1^2 + z_1^2}$$

$$\leqslant \sum \frac{x_1^2}{x_1^5 + y_1^2 + z_1^2}.$$

So we only need to prove ① in the case when $xyz = 1$.
Since

$$\sum \frac{x^5}{x^5 + y^2 + z^2} = \sum \frac{x^5}{x^5 + xyz(y^2 + z^2)}$$

$$= \sum \frac{x^4}{x^4 + y^3 z + yz^3}$$

$$\geqslant \sum \frac{x^4}{x^4 + y^4 + z^4} = 1,$$

the left hand side of ① holds.

While

$$\sum \frac{x^2}{x^5 + y^2 + z^2} = \sum \frac{x^2 \cdot xyz}{x^5 + xyz(y^2 + z^2)}$$

$$= \sum \frac{x^2 yz}{x^4 + yz(y^2 + z^2)},$$

by the AM-GM inequality,

$$x^4 + x^4 + y^3 z + yz^3 \geq 4x^2 yz,$$

$$x^4 + y^3 z + y^3 z + y^2 z^2 \geq 4xy^2 z,$$

$$x^4 + yz^3 + yz^3 + y^2 z^2 \geq 4xyz^2,$$

$$y^3 z + yz^3 \geq 2y^2 z^2,$$

the summation of the above four inequalities leads to

$$x^4 + yz(y^2 + z^2) \geq x^2 yz + xy^2 z + xyz^2.$$

Therefore

$$\sum \frac{x^2}{x^5 + y^2 + z^2} = \sum \frac{x^2 yz}{x^4 + y^3 z + yz^3}$$

$$\leq \sum \frac{x^2 yz}{x^2 yz + xy^2 z + xyz^2} = 1.$$

This is just the right hand side of inequality ①.

Comments Boreico Iurie from Moldova won a special prize for his outstanding solution. Observe that

$$\frac{x^5 - x^2}{x^5 + y^2 + z^2} - \frac{x^5 - x^2}{x^3(x^2 + y^2 + z^2)}$$

$$= \frac{x^2(x^3 - 1)^2(y^2 + z^2)}{x^3(x^5 + y^2 + z^2)(x^2 + y^2 + z^2)} \geq 0.$$

Therefore

$$\sum \frac{x^5 - x^2}{x^5 + y^2 + z^2} \geq \sum \frac{x^5 - x^2}{x^3(x^2 + y^2 + z^2)}$$

$$= \frac{1}{x^2 + y^2 + z^2} \sum \left(x^2 - \frac{1}{x} \right)$$

$$\geqslant \frac{1}{x^2 + y^2 + z^2} \sum (x^2 - yz) \, (\text{because } xyz \geqslant 1)$$

$$\geqslant 0.$$

Second Day
9:30 – 13:30 July 14, 2005

Consider the sequence a_1, a_2, $\cdots$ defined by

$$a_n = 2^n + 3^n + 6^n - 1 (n = 1, 2, \cdots).$$

Determine all positive integers that are relatively prime to every term of the sequence. (proposed by Poland)

Solution (posed by Luo Ye)

First, we will prove the following result: for a fixed prime $p(p \geqslant 5)$,

$$2^{p-2} + 3^{p-2} + 6^{p-2} - 1 \equiv 0 \pmod{p}. \qquad \textcircled{1}$$

Since p is a prime no less than 5, $(2, p) = 1$, $(3, p) = 1$, and $(6, p) = 1$. By Fermat's little theorem, we have

$$2^{p-1} \equiv 1 \pmod{p}, 3^{p-1} \equiv 1 \pmod{p}, 6^{p-1} \equiv 1 \pmod{p}.$$

Therefore

$$3 \cdot 2^{p-1} + 2 \cdot 3^{p-1} + 6^{p-1} \equiv 3 + 2 + 1 = 6 \pmod{p},$$

i. e.

$$6 \cdot 2^{p-2} + 6 \cdot 3^{p-2} + 6 \cdot 6^{p-2} \equiv 6 \pmod{p}.$$

Simplifying gives

$$2^{p-2} + 3^{p-2} + 6^{p-2} - 1 \equiv 0 \pmod{p}.$$

So $\textcircled{1}$ holds, and $a_{p-2} = 2^{p-2} + 3^{p-2} + 6^{p-2} - 1 \equiv 0 \pmod{p}$.

It is trivial that $a_1 = 10$ and $a_2 = 48$.

For any integer n greater than 1, it has a prime factor p. If

$p \in \{2, 3\}$, then $(n, a_2) > 1$. If $p \geqslant 5$, then $(n, a_{p-2}) > 1$. Therefore we can claim that every integer greater than 1 does not match the condition.

Since 1 is co-prime to every other positive integer, 1 is the only number satisfying the condition.

⑤ Let $ABCD$ be a given convex quadrilateral with sides BC and AD equal in length and not parallel. Let E and F be interior points of the sides BC and AD respectively such that $BE = DF$. The lines AC and BD meet at P, the lines BD and EF meet at Q, the lines EF and AC meet at R. Consider all triangles PQR as E and F vary. Show that the circumcircles of these triangles have a common point other than P. (proposed by Poland)

Solution (posed by Zhao Tongyuan)

Since BC and AD are not parallel, the circumcircles of triangle APD and BPC are not tangent to each other. Otherwise, construct a common tangent line SS' through tangent point P, then

$$\angle DPS = \angle DAP, \angle BPS' = \angle BCP.$$

It follows from the equality $\angle DPS = \angle BPS'$ that $\angle DAP = \angle BCP$. Then $AD \parallel BC$, which is in contradiction with the condition.

Let the second common point of the circumcirles of triangles BCP and DAP be O, which is fixed. With loss of generality, let O be an interior point of triangle DPC. We will prove that the circumcirle of triangle PQR passes through O as E and F vary.

Connect the lines OA, OB, OC, OD, OE, OF, OP, OQ, OR, as show in the figure. Since B, C, O, P and O, P, A, D are concyclic, then

$$\angle OBC = \angle OPC, \angle OPC = \angle ADO \Rightarrow \angle OBC = \angle ADO.$$

Similarly, $\angle OCB = \angle DPO = \angle DAO$.

Together with $AD = BC$, we get $\triangle OBC \cong \triangle ODA$. So

$$OB = OD, \angle OBE = \angle ODF.$$

Note that $BE = DF$, so the triangles OBE and ODF are congruent, giving $OE = OF$ and $OB = OD$.

The equalities $\angle FOE = \angle FOB + \angle BOE = \angle BOF + \angle FOD = \angle BOD$ imply that the triangles BOD and FOE are similar. This means $\angle EFO = \angle BDO$, i.e. $\angle QFO = \angle QDO$, so the points Q, F, D, O are concyclic. Therefore

$$\angle RQO = \angle FDO.$$

Since O, P, A, D are concyclic, we have $\angle FDO = \angle ADO = \angle RPO$, so $\angle RQO = \angle RPO$. We conclude that the points O, R, P, and Q are concyclic, i.e. the circumcircle of PQR passes through O.

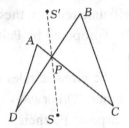 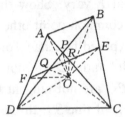

In a mathematical competition 6 problems were posed to the contestants. Each pair of problems was solved by more than $\frac{2}{5}$ of the contestants. Nobody solved all 6 problems. Show that there are at least 2 contestants who each solved exactly 5 problems. (proposed by Romania)

Solution (posed by Shao Xuancheng)

Assume there were n contestants C_1, C_2, $\cdots$, C_n, and the six problems were P_1, P_2, P_3, P_4, P_5, P_6.

Let $S = \{(C_k; P_i, P_j) \mid 1 \leqslant k \leqslant n, 1 \leqslant i < j \leqslant 6, C_k$ solved both P_i and $P_j\}$.

Now we will count $|S|$.

Let x_{ij} be the number of contestants having solved both P_i and P_j $(1 \leqslant i < j \leqslant 6)$. By hypothesis, $x_{ij} > \frac{2}{5}n \Leftrightarrow x_{ij} \geqslant \frac{2n+1}{5}$. Therefore

$$|S| = \sum_{1 \le i < j \le 6} x_{ij} \ge \sum_{1 \le i < j \le 6} \frac{2n+1}{5} = C_6^2 \cdot \frac{2n+1}{5} = 6n+3. \qquad \text{①}$$

If the number of contestants who solved exactly five problems was at most one, by hypothesis, no contestants has solved all problems, so the other contestants each solved four problems at most.

Let a_1, a_2, $\cdots$, a_n be the number of problems solved by contestants C_1, C_2, $\cdots$, C_n respectively. Without loss of generality, we can assume that $5 \ge a_1 \ge a_2 \ge \cdots \ge a_n \ge 0$.

If $4 \ge a_1$, then $4 \ge a_k (1 \le k \le 4)$,

$$|S| = \sum_{k=1}^{n} C_{a_k}^2 \le \sum_{k=1}^{n} C_4^2 = 6n,$$

this is in contradiction with ①. Therefore $a_1 = 5$, $4 \ge a_2 \ge \cdots \ge a_n$.

It is trivial that $n \ge 2$. If $a_n \le 3$, then

$$|S| = \sum_{k=1}^{n} C_{a_k}^2 \le C_5^2 + (n-2)C_4^2 + C_3^2 = 6n+1.$$

This is also in contradiction with ①. So $a_n \ge 4$, i.e. $a_2 = a_3 = \cdots = a_n = 4$.

Assume, without loss of generality, that the five problems C_1 solved were P_1, P_2, P_3, P_4, P_5, and the number of contestants who solved the problem $P_j (1 \le j \le 6)$ was $b_j (1 \le j \le 6)$. Then

$$b_1 + b_2 + \cdots + b_6 = a_1 + a_2 + \cdots + a_n$$

$$= 5 + 4(n-1) = 4n+1. \qquad \text{②}$$

Consider

$$\sum_{j=2}^{6} x_{1j} \ge \sum_{j=2}^{6} \frac{2n+1}{5} = 2n+1. \qquad \text{③}$$

Since the problem P_1 has been solved by b_1 contestants, and one of them solved five problems and the others solved four problems,

$$\sum_{j=2}^{6} x_{1j} = 4 + 3(b_1 - 1) = 2b_1 + 1. \qquad \text{④}$$

Combining ③ and ④ we obtain

$$2n+1 \leqslant 3b_1 +1,$$

which can be rewritten as $b_1 \geqslant \dfrac{2n}{3}$.

Similarly, $b_k \geqslant \dfrac{2n}{3}$, for $k = 1, 2, 3, 4, 5.$ ⑤

Consider

$$\sum_{j=1}^{6} x_{j6} \geqslant \sum_{j=1}^{5} \frac{2n+1}{5} = 2n+1,$$ ⑥

By a similar argument, since exactly b_6 contestants solved P_6, and each of these contestants solved four problems, we have

$$\sum_{j=1}^{6} x_{j6} = 3b_6.$$ ⑦

Together with ⑥ and ⑦ we obtain

$$3b_6 \geqslant 2n+1,$$

which we can rewrite as $b_6 \geqslant \dfrac{2n+1}{3}.$ ⑧

If $n \not\equiv 0 \pmod 3$, then $\dfrac{2n}{3}$ is not an integer. By ⑤ we obtain

$$b_k > \frac{2n}{3} (1 \leqslant k \leqslant 5),$$

which implies

$$b_k \geqslant \frac{2n+1}{3} (1 \leqslant k \leqslant 5).$$ ⑨

Together with ⑧ and ⑨ we obtain

$$b_1 + b_2 + \cdots + b_6 \geqslant 6 \cdot \frac{2n+1}{3} = 4n+2,$$

which is in contradiction with ②.

Therefore $n \equiv 0 \pmod 3$, and ⑧ implies $b_6 > \dfrac{2n}{3}$. Since b_6 and $\dfrac{2n}{3}$ are both integers,

$$b_6 \geqslant \frac{2n}{3} + 1. \tag{⑩}$$

Together with ⑤ and ⑩ we obtain

$$b_1 + b_2 + \cdots + b_6 \geqslant 5 \cdot \frac{2n}{3} + \frac{2n}{3} + 1 = 4n + 1. \tag{⑪}$$

Compare with ②, the ⑪ should be an equality, so are ⑤ and ⑩. Hence

$$b_6 = \frac{2n}{3} + 1.$$

Consider

$$\sum_{1 \leqslant i < j \leqslant 5} x_{ij} \geqslant \sum_{1 \leqslant i < j \leqslant 5} \frac{2n+1}{5} = 2(2n+1) = 4n+2. \tag{⑫}$$

Since one contestant solved P_1, P_2, P_3, P_4, P_5, whereas b_6 contestants solved exactly three problems of P_1, P_2, P_3, P_4, P_5, and $n-1-b_6$ contestants solved exactly four problems of P_1, P_2, P_3, P_4, P_5, we have

$$\sum_{1 \leqslant i < j \leqslant 5} x_{ij} = C_5^2 + b_6 \cdot C_3^2 + (n-1-b_6)C_4^2$$
$$= 6n + 4 - 3b_6 = 4n + 1,$$

which is in contradiction with ⑫. This completes the proof.

2006 (Ljubljana, Slovenia)

The 47th IMO (International Mathematical Olympiad) was hosted by Slovenia in Ljubljana during July 6 - 18 in 2006.

The leader of Chinese IMD 2006 team was Li Shenghong who was from Zhejiang University and deputy leader was Leng Gangsong who was from Shanghai University. The Chinese IMO2006 team came first among the nations with 6 golds. Here are the results of the six contestants:

Liu Zhiyu	No. 1 Middle School Attached to Central China Normal University	42 points	gold medal
Shen Caili	Zhejiang Zhenhai High School	37 points	gold medal
Deng Yu	Shenzhen Senior High School	35 points	gold medal
Jin Long	the Affiliated High School of Northeast Normal University	35 points	gold medal
Ren Qingchun	Yaohua Middle School , Tianjin	34 points	gold medal
Gan Wenyin	No. 3 High School of WISCO	31 points	gold medal

First Day
9:00 – 13:30 July 12, 2006

Let ABC be a triangle with incentre I. A point P in the interior of the triangle satisfies

$$\angle PBA + \angle PCA = \angle PBC + \angle PCB.$$

Show that $AP \geqslant AI$, and that the equality holds if and only if $P = I$.

Solution   Since $\angle PBA + \angle PCA = \angle PBC + \angle PCB$, we get

$$2(\angle PBC + \angle PCB) = \angle PBC + \angle PCB + \angle PBA + \angle PCA$$

$$= \angle ABC + \angle ACB,$$

i.e. $\angle PBC + \angle PCB = \dfrac{1}{2}(\angle ABC + \angle ACB) = 90° - \dfrac{1}{2}\angle BAC$,

$\angle BPC = 90° + \dfrac{1}{2}\angle BAC$.

On the other hand $\angle BIC = 90° + \dfrac{1}{2}\angle BAC$. Hence $\angle BPC =$

$\angle BIC$, B, C, I, P are concyclic.

It is easy to show that the middle point M of arc BC is the center of circumcircle of triangle BIC. This is because M is also the point where AI intersects the circumcircle of triangle ABC. Furthermore,

$$\angle BIM = \frac{1}{2}(\angle A + \angle B),$$

$$\angle MBI = \frac{1}{2}\angle B + \angle MBC = \frac{1}{2}(\angle B + \angle A),$$

so $$MI = MB.$$

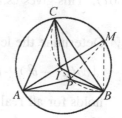

From triangle APM, $AP + PM \geqslant AM = AI + IM = AI + MP$.

Therefore $AP \geqslant AI$. The equality holds if and only if P lies on the line segment AI, that is $I = P$.

② Let P be a regular 2006-gon. A diagonal of P is said to be *good* if its endpoints divide the perimeter of P into two parts, each consists of an odd number of the sides of P. The sides of P are by definition *good*.

Suppose P is partitioned into triangles by 2003 diagonals, no two of which have a common point in the interior of P. Find the maximum number of isosceles triangles having two good sides that could appear in such a configuration.

Solution Let ABC be a triangle in the partition. Here are AB denotes the part of the perimeter of P outside the triangle and between points A and B, and similarly for arcBC and arcCA. Let a, b and c be the number of sides on arcAB, arcBC, arc CA respectively. Note that $a + b + c = 2006$. By parity check, if an isosceles triangle having two good sides, these sides must be two equal sides.

We call such isosceles triangles *good*.

Let one of the *good* triangles be ABC with $AB = AC$, Also, inscribe our polygon into a circle.

If there is another *good* triangle in arcAB, the two equal good sides cut off an even number of sides in arcAB. Since there is an odd number of sides in arcAB, there must be one side not belonging to any good triangle. The same holds for arcAC.

So every good triangle corresponds to at least two sides of P. Hence there are no more than $2\,006/2 = 1\,003$ *good* triangles.

And this bound can be achieved. Let $P = A_1 A_2 A_3 \cdots A_{2\,006}$. Draw diagonals between $A_1 A_{2k+1}$ ($1 \leqslant k \leqslant 1\,002$) and $A_{2k+1} A_{2k+3}$ ($1 \leqslant k \leqslant 1\,001$). This gives us the required $1\,003$ *good* triangles.

🔷 Determine the least real number M such that the inequality

$$|ab(a^2 - b^2) + bc(b^2 - c^2) + ca(c^2 - a^2)| \leqslant M(a^2 + b^2 + c^2)^2$$

holds for all real numbers a, b and c.

Solution   Factorizes the left side of the above inequality and the problem is reduced to finding the smallest number M that satisfies the inequality

$$|(a-b)(b-c)(c-a)(a+b+c)| \leqslant M(a^2 + b^2 + c^2)^2.$$

Let $x = (a-b)$, $y = (b-c)$, $z = (c-a)$, and $s = (a+b+c)$. Then the inequality becomes

$$|xyzs| \leqslant \frac{M}{9}(x^2 + y^2 + z^2 + s^2)^2$$

with the property that $x + y + z = 0$.

Since $x + y + z = 0$, without loss of generality, we can suppose x, y to have the same sign and both positive (otherwise we can replace a, b, c by $-a$, $-b$, $-c$).

Now, for any fixed $x + y = 2m$, let $x = y = m$, so $z = -2m$, the left side gets greater and the right gets smaller and the inequality still holds.

$$|xyzs| \leqslant \left| \left(\frac{x+y}{2} \right)^2 zs \right| = |2m^3 s|,$$

$$(x^2 + y^2 + z^2 + s^2)^2 \geqslant \left(\frac{(x+y)^2}{2} + z^2 + s^2 \right)^2$$

$$= (6m^2 + s^2)^2,$$

By the AM-GM inequality,

$$(6m^2 + s^2)^2 = (2m^2 + 2m^2 + 2m^2 + s^2)^2$$

$$\geqslant (4\sqrt[4]{(2m^2)^3 s^2})^2 = 16\sqrt{2}\,|m^3 s|,$$

so
$$\frac{16\sqrt{2}M}{9}|m^3 s| \geqslant |2m^3 s|$$

i.e. $M \geqslant \dfrac{9\sqrt{2}}{16}$.

The conditions for the equality can now be stated as $x = y$, $2m^2 = s^2$, or restated as $2b = a + c$, $(c-a)^2 = 18b^2$. Setting $b = 1$ yields $a = 1 - \dfrac{3}{2}\sqrt{2}$, $c = 1 + \dfrac{3}{2}\sqrt{2}$.

We can conclude that $M = \dfrac{9\sqrt{2}}{16}$ is indeed the smallest constant satisfying the inequality, with the equality for any triple (a, b, c) proportional to $\left(1 - \dfrac{3}{2}\sqrt{2}, 1, 1 + \dfrac{3}{2}\sqrt{2} \right)$ up to permutation.

Second Day
9:00 – 13:30 July 13, 2006

Determine all pairs (x, y) of integers such that

$$1 + 2^x + 2^{2x+1} = y^2.$$

Solution   It is easy to show that $x \geqslant 0$. Since $(-y)^2 = y^2$, we only need to find all solutions with $y > 0$.

If $x = 0$, then $y = \pm 2$;

Now y is odd, let $y = 2n + 1$. Then $2^x(2^{x+1} + 1) = 4n(n+1)$. It is clear that there is no solution for $x = 1, 2$.

Assume $x > 2$. Since $(2^x, 2^{x+1} + 1) = 1$, $(n, n+1) = 1$, $\dfrac{4n}{2^x}$ or

$\dfrac{4(n+1)}{2^x}$ are integers.

Case 1: Let $4n = a \cdot 2^x$, then $a^2 \cdot 2^x + 4a = 8 \cdot 2^x + 4$. It yields $2^x = \dfrac{4(1-a)}{a^2 - 8} \geqslant 8$, so $a = 1, 2$. In both cases we get a contradiction.

Case 2: Let $4(n+1) = a \cdot 2^x$, then $a^2 \cdot 2^x - 4a = 8 \cdot 2^x + 4$. It yields $2^x = \dfrac{4(1+a)}{a^2 - 8} \geqslant 8$, so $a = 1, 2$, or 3. It is easy to check that only $a = 3$ is good. So $x = 4$, and $y^2 = 529$.

Thus we have the complete list of solutions (x, y): $(0, 2)$, $(0, -2)$, $(4, 23)$, $(4, -23)$.

⑤ Let $P(x)$ be a polynomial of degree $n > 1$ with integer coefficients and let k be a positive integer. Consider the polynomial $Q(x) = P(P(\cdots P(P(x)) \cdots))$, where P occurs k times. Prove that there are at most n integers t such that $Q(t) = t$.

Solution The claim is obvious if every integer fixed point of Q is a fixed point of P itself. In the sequel, assume that this is not the case. Take any integer x_0 such that $Q(x_0) = x_0$ and $P(x_0) \neq x_0$. Define inductively $x_{i+1} = P(x_i)$ for $i = 0, 1, 2, \cdots$, then $x_k = x_0$.

It is evident that

$$u - v \mid P(u) - P(v), \text{ for distinct integers } u, v. \qquad (1)$$

(Indeed, if $P(x) = \sum a_i x^i$ then each $u - v \mid a_i(u^i - v^i)$.) Therefore each term in the chain of (nonzero) differences

$$x_0 - x_1, x_1 - x_2, \cdots, x_{k-1} - x_k, x_k - x_{k+1}, \qquad (2)$$

is a divisor of the next one; and since $x_k - x_{k+1} = x_0 - x_1$, all these differences have equal absolute values. Take $x_m = \min(x_1, \cdots, x_k)$, this means that $x_{m-1} - x_m = -(x_m - x_{m+1})$. Thus $x_{m-1} = x_{m+1}$ ($\neq x_m$).

It follows that consecutive differences in the sequence (2) have opposite signs. Consequently, x_0, x_1, x_2, $\cdots$ is an alternating sequence of two distinct values. In other words, every integer fixed point of Q is a fixed point of the polynomial $P(P(x))$. Our task is to prove that there are at most n such points.

Let a be one of them so that $b = P(a) \neq a$ (we have assumed that such an a exists). Then $a = P(b)$. Take any other integer fixed point α of $P(P(x))$ and let $P(\alpha) = \beta$, so that $P(\beta) = \alpha$. The numbers α and β need not be distinct (α can be a fixed point of P), but each of α, β is different from each of a, b. Applying property (1) to the four pairs of integers (α, a), (β, b), (α, b), (β, a), we get that the numbers $\alpha - a$ and $\beta - b$ divide each other, and also $\alpha - b$ and $\beta - a$ divide each other. Consequently

$$\alpha - b = \pm (\beta - a), \quad \alpha - a = \pm (\beta - b). \tag{3}$$

Suppose we have a plus sign in both instances: $\alpha - b = \beta - a$ and $\alpha - a = \beta - b$. Subtracting yields $a - b = b - a$, a contradiction, as $a \neq b$. Therefore at least one equality in (3) holds with a minus sign. This means that $\alpha + \beta = a + b$; equivalently $a + b - \alpha - P(\alpha) = 0$.

Denote $a + b$ by C. We have shown that every integer fixed point of Q other that a and b is a root of the polynomial $F(x) = C - x - P(x)$. This is of course true for a and b as well. Since P has degree $n > 1$, the polynomial F has the same degree. So it cannot have more than n roots. Hence the result.

⬤ Assign to each side b of a convex polygon P the maximum area of a triangle that has b as a side and is contained in P. Show that the sum of the areas assigned to the sides of P is at least twice the area of P.

Solution *Every convex $(2n)$-gon, of area S, has a side and a vertex that jointly span a triangle of area not less than S/n.*

Proof of the lemma

By *main diagonals* of the $(2n)$-gon we shall mean those which partition the $(2n)$-gon into two polygons with equally many sides. For any side b of the $(2n)$-gon, denote by Δ_b the triangle ABP where A, B are the endpoints of b and P is the intersection point of the main diagonals AA', BB'. We claim that the union of triangles Δ_b, taken over all sides, covers the whole polygon.

To show this, choose any side AB and consider the main diagonal AA' as a directed line segment. Let X be any point in the polygon, not on any main diagonal. For definiteness, let X lie on the left side of the ray AA'. Consider the sequence of main diagonals AA', BB', CC', $\cdots$, where A, B, C, $\cdots$ are consecutive vertices, situated to the right of AA'.

The n-th item in this sequence is the diagonal $A'A$ (i. e. AA' reversed), having X on its right side. So there are two successive vertices K, L in the sequence A, B, C, $\cdots$ before A' such that X still lies to the left of KK' but to the right of LL'. This means that X is in the triangle $\Delta_{l'}$, where $l' = K'L'$. We can apply the analogous reasoning to points X on the right of AA' (points lying on the main diagonals can be safely ignored). Thus indeed the triangles Δ_b for all b jointly cover the whole polygon.

The sum of their areas is no less than S. So we can find two opposite sides, say $b = AB$ and $b' = A'B'$ (with AA', BB' being main diagonals) such that $[\Delta_b] + [\Delta_{b'}] \geqslant S/n$, where $[\cdots]$ stands for the area of a region. Let AA' and BB' intersect at P. Assume without loss of generality that $PB \geqslant PB'$.

Then

$$[ABA'] = [ABP] + [PBA'] \geqslant [ABP] + [PA'B']$$

$$= [\Delta_b] + [\Delta b'] \geqslant S/n,$$

proving the lemma.

Now, let P be any convex polygon, of area S, with m sides a_1, $\cdots$, a_m. Let S_i be the area of the greatest triangle in P with side a_i. Suppose, contrary to the assertion, that

$$\sum_{i=1}^{m} \frac{S_i}{S} < 2.$$

Then there exist rational numbers $q_1, \cdots, q_m$ such that $\sum q_i = 2$ and $q_i > S_i/S$ for each i.

Let n be a common denominator of the m fractions $q_1, \cdots, q_m$. Write $q_i = k_i/n$; so $\sum k_i = 2n$. Partition each side a_i of P into k_i equal segments, creating a convex $(2n)$-gon of area S (with some angles of size $180°$), to which we apply the lemma. Accordingly, this induced polygon has a side b and a vertex H spanning a triangle T of area $[T] \geqslant S/n$. If b is a portion of a side a_i of P, then the triangle W with base a_i and summit H has area

$$[W] = k_i \cdot [T] \geqslant k_i \cdot S/n = q_i \cdot S > S_i,$$

in contradiction with the definition of S_i. This completes the proof.